KB039796

Digital SAT Math for 800

Digital SAT Math for 800

발　행　2022년 11월 25일 초판 1쇄
　　　　2023년 04월 20일 초판 2쇄
저　자　기혜연
발행인　최영민
발행처　헤르몬하우스
주　소　경기도 파주시 신촌로 16
전　화　031 – 8071 – 0088
팩　스　031 – 942 – 8688
전자우편　hermonh@naver.com
출판등록　2015년 3월 27일
등록번호　제406 – 2015 – 31호

ISBN　　979 – 11 – 92520 – 15 – 5　(53410)

✜ 저자직강 인터넷 강의는 SAT, AP No.1 인터넷 강의 사이트인 마스터프랩 (www.masterprep.net) 에서 보실 수 있습니다.

Liz's Mathism

DIGITAL SAT MATH FOR

SAT Math Concepts from A to Z and 4 practice tests

기혜연 지음(Liz Ki)

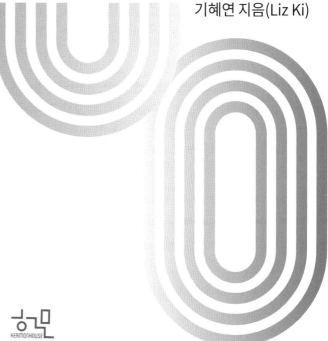

HERMONHOUSE

Preface

대학에서 Mathematics education을 전공하고, 미국 Virginia Fairfax County Public School 그리고 이후 International Schools in South Korea, 압구정 유학전문학원에서 일하면서 일관되게 해결하고 싶은 문제는 '학생들이 어떻게 하면 수학을 쉽게 할 수 있고 즐길 수 있을까' 이다.

Problem-solving에는 여러 가지 방식이 존재하지만, Standardized Test Prep을 위한 문제풀이, 그리고 Middle School and High School 때 배우는 수학은 분명 제일 쉽고 정확한 방식(One Best Way To Solve)이 있을 것이라는 것이 나의 철학이다.

좋은 수학 교사는 개념을 가르치는 것을 넘어서, 최적의 문제풀이법(Problem Solving Methods)을 알려주어야 하고, Algebra 2에서 배웠던 개념이 Pre-calculus에서도 똑같이 적용되어야 한다고 생각한다.

Mathematics는 STEM 분야에서 기본적인 언어일 뿐이다. STEM 분야의 전공들이 점점 더 많아지는 시대에 Mathematics는 잘하는 친구들만 제대로 배워야 하는 과목이 아니라, 이제는 잘하지 않아도 잘 배워야 하는 과목이 되었다.

10년 이상 수학교육의 길을 걸으면서 American Mathematics Series의 각 과목의 문제풀이법, ACT Math, Digital SAT Math의 문제풀이법에 대해 확실한 해답을 찾아서 이제야 책을 내게 되었다.

Digital SAT Math Section은 오답이 하나도 없어야 만점이 가능한 시험이기에, 우선 학생들이 왜 오답을 내는지, 어떤 문제에서 주로 오답을 내는지 분석해야 했다.

거부감 없이 자신들의 약점을 드러내어 오답 정리를 꼭 해야 하는 문제들을 알게 해준 나의 학생들과 성실하게 잘 따라와서, 800점을 맞아준 많은 학생들에게 감사함을 전하고 싶다.

모든 일에 함께하시는 나의 주님께 이 모든 영광을 올린다. 하나의 아이디어가 교재로 제작될 수 있기까지 도와주신 마스터프렙 권주근 대표님과 매년 발전하도록 도와주시는 메이커스 어학원 정병철 대표님께도 감사를 말을 전한다.

Liz Ki

저자소개

Liz Ki 선생님은 대학교에서 수학교육을 전공하여 탄탄한 수학이론과 전문적인 수학 교수법을 익혔고 이후 미국 버지니아 공립학교 및 한국 국제학교 교사로서 일하였다.

현재는 서울 강남 압구정 SAT, AP 유학전문학원에서 미국 명문 보딩스쿨 학생들과 국내 외국인 학교 및 국제학교 학생들을 가르치면서 수학교재 집필에 힘쓰고 있다.

한국어와 영어에 능통하고 동시에 한국 수학 및 국제학교 수학 과정 모두에 정통한 수준 높은 수학 강의로 정평이 나 있으며, AP Calculus와 SAT 만점자들의 강사이자 만점이 보장되는 선생님으로서 유명하다.

유학생 대상 인터넷 강의 1위인 마스터프렙(www.masterprep.net)과 압구정 메이커스 어학원(https://makersaca.com)에서 현장강의를 하면서 학생들과 활발한 소통을 하고 있다.

Digital SAT Math Section 개요

Characteristic	Digital SAT Suite Math Section
Administration	Two-stage adaptive testing
	1st Module: Average level of problem set
	2nd Module: Easy or Advanced level of problem set
Test length	1st Module : 20 operational questions and 2 pretest questions
	2nd Module: 20 operational questions and 2 pretest questions
Time per module	1st Module: 35 minutes
	2nd Module: 35 minutes
	Total: 70 minutes
Score reported	Section score (constitutes half of total score)
	SAT :200-800
Questions by Content Domain	1. Algebra: 35%
	2. Advanced Math: 35%
	3. Problem-solving and Data analysis: 15%
	4. Geometry and Trigonometry: 15%

Digital SAT Math for 800 공부법

■ Algebra I and II Topics가 중요하다.

Digital SAT Math 에 나오는 문제 중에서 Algebra 1 and 2 와 관련된 문제는 70% 정도이다. Algebra Topics 중에서도 Linear function, Quadratic function, Exponential function, System of equations, Polynomial function 을 잘하는 것이 관건이다.

Algebra 를 먼저 완벽하게 한 다음, 다른 주제를 공부하는 것을 추천한다.

■ SAT is a standardized test.

SAT 는 College Board 가 제작한 Standardized Test 임을 기억해야 한다. 시험에 나오는 문제와 주제가 정해져 있다. 시험에 나오는 문제만 공부하면 된다. Algebra 가 중요하다고 해서, Algebra 1 and 2 책을 다시 펼칠 필요 없다. 이 책에 나오는 주제를 위주로 공부하면 된다. 또한, 책에 적힌 문제풀이 방식대로 풀어야 어려운 문제도 쉽게 풀 수 있다.

■ 오답 정리가 만점의 비밀이다.

자신이 틀리는 문제는 시간이 지나서 또 풀어도 틀릴 가능성이 거의 100% 이다. 한번 틀렸던 문제는 맞을 때까지 풀어야 한다. 새로운 문제를 하나 더 푸는 것이 풀었던 문제를 반복하는 것보다 즐거울 수 있다. 하지만, 오답을 낸 문제를 반복해서 푸는 친구가 만점을 받을 것이다.

■ Pay attention to details.

수학 개념을 알아도 틀리는 것이 SAT 문제이다. Details를 잘 확인하면서 읽어야 한다. 당연히, 하루 만에 Details를 신경 쓰면서 읽는 것은 가능하지 않다. Details를 고려하는 습관을 길러야 한다. 시간이 걸리겠지만, 세심하게 읽는 습관이 앞으로의 삶에서 성공적인 학업과 일을 하는 데 큰 도움이 될 것이라 생각한다. 그런 면에서 노력해서 College Board SAT 고득점을 맞은 친구는 앞으로 어떤 일을 하더라도 잘 해낼 것이라고 믿는다.

Contents

Chapter **1**

Linear Function and Exponential Function

1.1 Linear Function

Definition 1: Function

Function is a relation for which each x has exactly one y.

By using the vertical line test, if the curve does not intersect more than once with the line, it is a function.

The x-intercept is $(x,0)$ or $f(x) = 0$

The y-intercept is $(0,y)$ or $f(0) = y$

Linear functions

3 forms of linear functions

1. standard form : $ax + by = c$, where x-intercept is $(\frac{c}{a}, 0)$ and y-intercept is $(0, \frac{c}{b})$.

2. slope-intercept form : $y = mx + b$, where m is a slope and b is a y-intercept.

3. point-slope form : $y - y_1 = m(x - x_1)$, where (x_1, y_1) is a point on the line.

Definition 2: Parallel lines and Perpendicular lines

If 2 lines are parallel, 2 lines have same slopes.

If 2 lines are perpendicular, slopes of 2 lines are opposite reciprocals.

1.1.1 Example

In the xy-plane below, line m is perpendicular to line l, which is not shown. Which of the following could be an equation of line l?

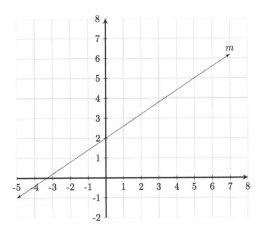

(A) $5x + 3y + 2 = 0$
(B) $5x - 3y + 2 = 0$
(C) $3x - 5y + 6 = 0$
(D) $3x + 5y - 6 = 0$

1.1.2 Example

$$p = 14.7 + 0.439d$$

The equation above shows the pressure p, in pounds per square inch, exerted on a diver at a depth of d feet(ft) below the surface of the water. Based on the equation, which of the following must be true?

I. A depth increase of 1000 feet is equivalent to a pressure increase of 439 pounds per square inch.

II. A depth increase of $\frac{1}{0.439}$ feet is equivalent to a pressure increase of 1 pound per square inch.

III. A depth increase of 1 feet is equivalent to a pressure increase of $\frac{1}{0.439}$ pound per square inch.

(A) I only
(B) II only
(C) I and III only
(D) I and II only

1.2 Exponential Function

Exponential function

$$f(x) = a \cdot (b)^x$$

where

$$a \neq 0, b \neq 1, \ and \ b > 0$$

a represents the initial value or y-intercept; $(0, a)$

When $a > 0$, b represents a growth factor if $b > 1$ or a decay factor if $0 < b < 1$

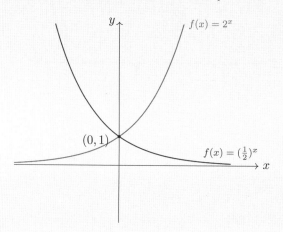

1.2.1 Example

The graph of an exponential function is shown in the xy-plane.

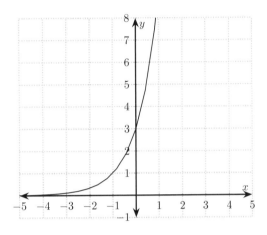

Which of the following could be an equation of the graph shown, where k is a positive constant?

(A) $y = 3^x + k$
(B) $y = 3^{x+k}$
(C) $y = 3^x - k$
(D) $y = 3^{x-k}$

1.2.2 Example

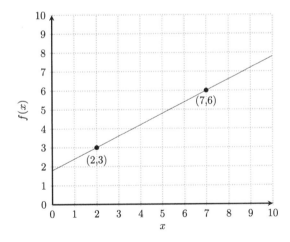

In the xy-plane, the graph of a linear function is shown. The graph of an exponential function (not shown) intersects the graph of the linear equation at $(2,3)$ and $(7,6)$. The point (r, s) is on the graph of the linear function and the point (r, t) is on the graph of the exponential equation, where $0 < r < 10$ and $s > t$, which of the following must be true?

(A) $0 < r < 2$
(B) $2 < r < 7$
(C) $7 < r < 10$
(D) $3 < r < 6$

1.3 Exponential Modeling

Exponential Modeling

$$f(t) = a \cdot (b)^{\frac{t}{k}}$$

a = Initial value; y-intercept
b = Decay factor $(0 < b < 1)$ or Growth factor $(b > 1)$
t = Time
k = Decay or Growth period

1.3.1 Example

$$f(t) = 50,000(1.1)^{\frac{t}{20}}$$

The function f, defined by the given equation, models population of a community t years after 2013 and the population was projected to increase by 10 percent every 20 years. Under these conditions, which of the following expression models the population of the community d decades after 2013? (1 decade=10 years)

(A) $f(t) = 50,000(1.1)^{2d}$
(B) $f(t) = 50,000(1.1)^{\frac{d}{2}}$
(C) $f(t) = 50,000(1 + \frac{0.1}{10})^{d}$
(D) $f(t) = 50,000(1 + \frac{0.1}{10})^{2d}$

1.4 System of Linear Equations

system of linear equations

1. Standard form of linear equations

$$\begin{cases} ax + by = p \\ cx + dy = q \end{cases}$$

(a) One solution

$$\frac{a}{c} \neq \frac{b}{d}$$

(b) No solution

$$\frac{a}{c} = \frac{b}{d} \neq \frac{p}{q}$$

(c) Infinitely many solutions

$$\frac{a}{c} = \frac{b}{d} = \frac{p}{q}$$

2. Slope-intercept form of linear equations

$$ax + b = cx + d$$

(a) One solution

$$a \neq c$$

(b) No solution

$$a = c \text{ but } b \neq d$$

(c) Infinitely many solutions

$$a = c \text{ and } b = d$$

1.4.1 Example

The system has no solution. Which of the following linear equations could be?

$$\text{I. } 5(x - 2) = 5x + 10$$
$$\text{II. } 5(x - 2) = 4x - 10$$

(A) I only
(B) II only
(C) I and II
(D) Neither I nor II

1.4.2 Example

$$\begin{cases} 2x - 3y = 1 \\ ky + 10x = m \end{cases}$$

In the system of equations above, k and m are constants. If the system has infinitely many solutions, what is the value of $\frac{k}{m}$?

(A) -3
(B) -2
(C) $-\frac{1}{3}$
(D) $-\frac{1}{2}$

1.5 Inequalities

Definition 3: System of inequalities

A solution of a system of inequalities is an ordered pair that is a solution of each inequality in the system. Once you graph each inequality in the system, the region that is common to all the graphs of the inequalities is the solution set.

$$\begin{cases} x + y < 3 \\ -x + y \geq -4 \end{cases}$$

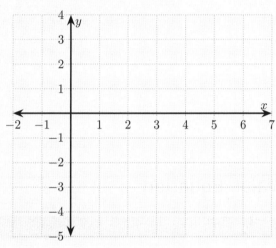

Inequalities			
$x > 4$	$x \geq 4$	$x < 4$	$x \leq 4$
x is greater than 4	x is at least 4	x is less than 4	x is at most 4
x is over 4	x is no less than 4		x is no more than 4

1.5.1   Example

$$0 \leq x \leq y \leq 8$$

What is the area, in the xy-plane, of the region consisting of all points that satisfy the inequality above?

1.5.2 Example

A science teacher arranges lab tables and chairs in a classroom, and a table can match with no more than 8 chairs. The classroom can hold a maximum of 54 tables and chairs. If t represents the number of tables and c represents the number of chairs, which of the following system of inequalities best describes the possible numbers of tables and chairs that can be used in the classroom?

(A) $\begin{cases} t + c \geq 54 \\ t \leq 8c \end{cases}$

(B) $\begin{cases} t + c \geq 54 \\ t \geq 8c \end{cases}$

(C) $\begin{cases} t + c \leq 54 \\ t \leq \frac{c}{8} \end{cases}$

(D) $\begin{cases} t + c \leq 54 \\ t \geq \frac{c}{8} \end{cases}$

1.6 Absolute Value Equation

Definition 4: Absolute Value Equation

$$| x | = a$$

1. If $a = 0$, $x = 0$

2. If $a > 0$, $x = -a$ or $x = a$

3. If $a < 0$, No solution

1.6.1 Example

$$|\frac{1}{2}x + 1| = 3$$

What is the sum of the solutions to the given equation?
(A) -4
(B) -1
(C) 0
(D) 4

Explanation

1.1.1 Example

The slope of line is $\frac{3}{5}$.

The slope of line l is a opposite reciprocal of $\frac{3}{5}$, which is $-\frac{5}{3}$.

Answer: (A)

1.1.2 Example

I. Since $\Delta p = 439, \Delta d = 1000$, slope is equal to $\frac{\Delta p}{\Delta d} = \frac{439}{1000} = 0.439$. This statement is true.

II. Since $\Delta p = 1, \Delta d = \frac{1}{0.439}$, slope is equal to $\frac{\Delta p}{\Delta d} = \frac{1}{\frac{1}{0.439}} = 0.439$. This statement is true.

III. Since $\Delta p = \frac{1}{0.439}, \Delta d = 1$, slope is equal to $\frac{\Delta p}{\Delta d} = \frac{1}{0.439}$. This statement is false.

Answer: (D)

1.2.1 Example

The horizontal asymptote is $y = 0$. The y-intercept of the graph is $(0, 3)$.

For (B), y-intercept is $(0, 3^k)$, where k is a positive constant and horizontal asymptote is $y = 0$.

Answer: (B)

1.2.2 Example

When a graph of exponential function, that passes through $(2, 3)$ and $(7, 5)$ is drawn on the plane, the graph of linear function is above the graph of exponential function where $2 < r < 7$.

Answer: (B)

1.3.1 Example

The population in 2013 is $50,000$. The growth factor is 1.1 and the growth period is 2 decades.

$f(t) = 50,000(1.1)^{\frac{d}{2}}$

Answer: (B)

1.4.1 Example

I. $5x - 10 = 5x + 10$. The system has no solution.

II. $5x - 10 = 4x - 10$. The system has one solutoin.

Answer: (A)

1.4.2 Example

$\frac{2}{10} = \frac{-3}{k} = \frac{1}{m}$

$\frac{-3}{k} = \frac{1}{m}$

$\frac{k}{m} = -3$

Answer: (A)

1.5.1 Example

$0 \le x \le 8, 0 \le y \le 8$, and $x \le y$

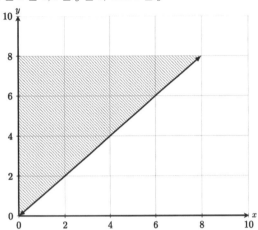

Answer: 32

1.5.2 Example

The maximum number of tables and chairs is 54. Thus, $t + c \le 54$.

If a table matches with chairs less than 8, more tables will be needed.

Then, $t \ge \frac{c}{8}$

Answer: (D)

1.6.1 Example

$\frac{1}{2}x + 1 = \pm 3$ $x = 4, -8$ The sum of the solutions is -4.

Answer: (A)

1.7 Chapter 1 from A to Z Problems

1.
$$5x - 2 = (x + 2) - px$$

In the equation shown, p is a constant. If the equation has no solution, what is the value of p?

(A) -4

(B) -3

(C) 3

(D) 4

2.
$$D = \frac{F}{N - F}$$

The given equation relates the variables D, N and F. Which of the following equations gives F in terms of D and N?

(A) $F = \frac{N}{D-1}$

(B) $F = \frac{N}{D+1}$

(C) $F = \frac{DN}{D-1}$

(D) $F = \frac{DN}{D+1}$

3. The function f is defined by $f(x) = (2)(1.495)^x + 7$. What is the y-intercept of the graph of $y = f(x)$ in the xy-plane?

(A) $(7,0)$

(B) $(2,0)$

(C) $(0,9)$

(D) $(0,8)$

4. The boiling point of water above sea level is modeled by a decreasing linear function of time. Which of the following scenarios could describe this linear relationship?

(A) The boiling point of water drops approximately 1.84 °F for every increase of 1,000 feet above sea level.

(B) The boiling point of water decreases by 0.2 % with every increase of 1,000 feet above sea level.

(C) The boiling point of water becomes half of the previous boiling point every foot.

(D) The boiling point of water increases by 0.00184 °F every foot.

5. The taxi fare is $10 for the first mile, plus $5 for each additional mile. If m represents the number of miles the taxi has traveled, which of the following functions gives the cost $C(m)$, in dollars, of paying the taxi for m miles?

(A) $C(m) = 10m - 5$

(B) $C(m) = 10m + 5$

(C) $C(m) = 5m + 10$

(D) $C(m) = 5m + 5$

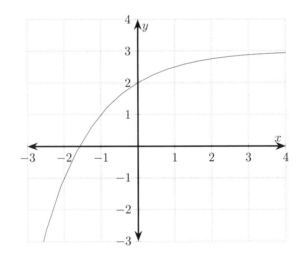

6. What is the equation of the graph shown?

(A) $y = -2^x + 3$

(B) $y = -2^{-x} + 3$

(C) $y = 2^x + 2$

(D) $y = -2^{-x} + 2$

7. In the xy-plane, line l has a slope of 4 and contains the point $(5,0)$. Line k is perpendicular to line l in the xy-plane. Which of the following could be an equation of line k?

(A) $4x - y = -20$

(B) $4x + y = -20$

(C) $x + 4y = 20$

(D) $x - 4y = 20$

8. The function $g(x)$ is defined by $g(x) = -\frac{4}{3}x - 8$. What is the graph of $y = g(x)$?

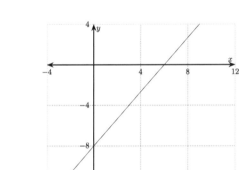

(A)

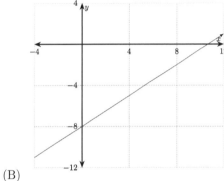

(B)

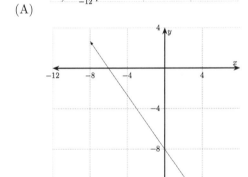

(C)

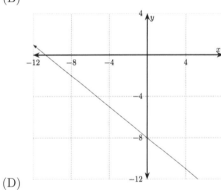

(D)

9. A group of geologists measured water movement through soil, called interflow, after exposing the soil to simulated rain. The interflow, $F(t)$, in cubic centimeters per minute(cm^3/min), after t minutes after water starts moving can be modeled by the function

$$F(t) = 60 \cdot \left(\frac{1}{2}\right)^{0.02t}$$

Which of the following describes the meaning of the fraction $\frac{1}{2}$ in the context described?

(A) The fraction of the 60 cm^3/min is the initial interflow

(B) The fraction of the 60 cm^3/min the soil had 0.02 minutes after the water started flowing

(C) The fraction of the 60 cm^3/min the soil had 50 minutes after the water started flowing

(D) The fraction of the 60 cm^3/min the soil had 72 minutes after the water started flowing

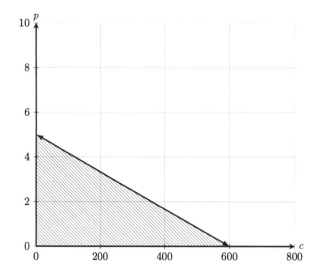

10. The shaded region shown models the possible pounds of aluminum cans, c, and tons of paper, p, that a school club is paid \$300 for recycling. Which of the following inequalities represents this relationship for $c \geq 0$ and $p \geq 0$?

(A) $0.5c + 60p \leq 300$

(B) $60c + 0.5p \leq 300$

(C) $5c + 60p \leq 300$

(D) $60c + 5p \leq 300$

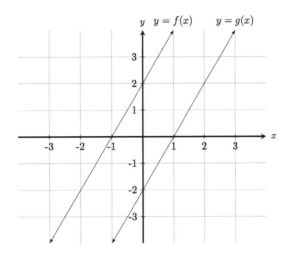

11. The graphs of the linear functions f and g are shown in the xy-plane above. Which of the following defines g in terms of f?

(A) $g(x) = f(x + 2)$

(B) $g(x) = f(x - 2)$

(C) $g(x) = f(x) + 2$

(D) $g(x) = f(x) - 2$

12. Computer repair technician charges an initial fee of \$50 and an additional \$20 fee for every $\frac{1}{2}$ hour of work. The technician visited a house and charged \$210. At this rate, in how many hours the technician charge for working on the computer in the house?

13. A company that manufactures calculators calculates its monthly profit by subtracting its manufacturing costs from its monthly revenue from sales. The equation

$$120x - 50x = 280,000$$

represents this situation for a month where x calculators are manufactured and sold. What is the meaning of $120x$ in this context?

(A) The total cost of manufacturing x calculators

(B) The total cost of manufacturing each calculator

(C) The total monthly revenue from each calculator sold

(D) The total monthly revenue from selling x calculators

14. Aidan invested a total of $5000 in two different bank accounts: Account A and Account B. The amount invested in the two different banks must differ by no more than $500. Which of the following systems represents all possible values for the amount of money a of Account A and the amount of money b of Account B ?

(A) $\begin{cases} a - b \le 500 \\ a + b = 5,000 \end{cases}$

(B) $\begin{cases} a - b = 500 \\ a + b \le 5,000 \end{cases}$

(C) $\begin{cases} -500 \le a - b \le 500 \\ a + b = 5,000 \end{cases}$

(D) $\begin{cases} -250 \le a - b \le 250 \\ a + b = 5,000 \end{cases}$

15. When Steven walks, he burns 4 calories per minute, and when he swims, he burns 9 calories per minute. If Steven spent 2 hours of swimming and burned between 1,600 and 1,608 calories, inclusive, on swimming and walking, what is the one possible value of integer minutes that he spent walking?

16.
$$\begin{cases} -7x + 3y = 3 \\ -14x + 6y = 3 \end{cases}$$

How many solutions does the given system of equation have?

(A) Zero

(B) Exactly one

(C) Exactly two

(D) Infinitely many

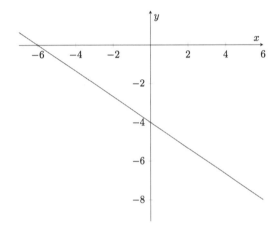

17. The graph shows the relationship between x and y. The relationship can be modeled by the equation $ax + by = -12$, where a and b are constants. What is the value of $\frac{a}{b}$?

18.

$$g(t) = 300,000(0.76)^{\frac{t}{12}}$$

The function g above can be used to determine a city's population for 10 years, and t is the number of months after the start of the 10 years. If the city's population decreases each year by $x\%$ over the prior year's population, what is the value of x?

19. The equation $s = 800d + 2,000,000$ approximates the age of the sediment, in years, at a depth of d feet below the top of the canyon. What is the increase in depth that is necessary to increase the age by 3224 years?

20. A certain colony of bacteria began with one cell, and the population triples every 20 minutes. What was the population of the colony after 2 hours?

21.

$$(x + 5) + (y - 3) = 43$$

$$(x + 5) - (y - 3) = 67$$

If (x, y) is the solution of the system of equations given above, what is the value of $8(x + 5)$?

Years	Number of capybaras
0	1870
10	2394
20	3064
30	3922
40	5021
50	6427
60	8228

22. The number of capybaras $c(t)$ living in a section of tropical forest in Brazil, where t is the number of years since 1960 is given in the table above. Which of the following best describes the relationship between time and the number of capybaras?

(A) It is linear because the number of capybaras is increasing by approximately the same factor every 10 years.

(B) It is linear because the number of capybaras is increasing by approximately the same number every 10 years.

(C) It is exponential because the number of capybaras is increasing by approximately the same factor every 10 years.

(D) It is exponential because the number of capybaras is increasing by approximately the same number every 10 years.

23. A certain LED lightbulb initially produces 3,300 lux of illumination. When the LED lightbulb is left on, its illumination decreases exponentially. Each minute for the next 70 minutes, the illumination decreased by approximately 2% from the previous minute. Which function f best models this situation, where $f(t)$ is the approximate lux of illumination after t minutes?

(A) $f(t) = 3,300(0.98)^{\frac{t}{70}}$

(B) $f(t) = 3,300(0.98)^{t}$

(C) $f(t) = 3,300(0.8)^{\frac{t}{70}}$

(D) $f(t) = 3,300(0.8)^{t}$

<table>
<thead>
<tr><th colspan="8">1.7 ANSWERS</th></tr>
<tr><th>Number</th><th>Answer</th><th>Number</th><th>Answer</th><th>Number</th><th>Answer</th><th>Number</th><th>Answer</th></tr>
</thead>
<tbody>
<tr><td>1</td><td>A</td><td>7</td><td>C</td><td>13</td><td>D</td><td>19</td><td>4.03</td></tr>
<tr><td>2</td><td>D</td><td>8</td><td>C</td><td>14</td><td>C</td><td>20</td><td>729</td></tr>
<tr><td>3</td><td>C</td><td>9</td><td>C</td><td>15</td><td>130,131,132</td><td>21</td><td>440</td></tr>
<tr><td>4</td><td>A</td><td>10</td><td>A</td><td>16</td><td>A</td><td>22</td><td>C</td></tr>
<tr><td>5</td><td>D</td><td>11</td><td>B</td><td>17</td><td>$\frac{2}{3}$</td><td>23</td><td>B</td></tr>
<tr><td>6</td><td>B</td><td>12</td><td>4</td><td>18</td><td>24</td><td></td><td></td></tr>
</tbody>
</table>

1.8 Explanation

1. **(A)**

   $5x - 2 = (1 - p)x + 2$

   $1 - p = 5$

   $p = -4$

2. **(D)**

   $D(N - F) = F$

   $DN = DF + F$

   $DN = (1 + D)F$

   $F = \frac{DN}{1+D}$

3. **(C)**

   $2(1.495)^0 + 7 = 9$

   $(0, 9)$

4. **(A)**

   Decreasing linear function has a negative average rate of change.

   (A) The average rate of change is -0.00184

   (B) The decay factor is 0.998

   (C) The decay factor is $\frac{1}{2}$.

   (D) The average rate of change is 0.00184

5. **(D)**

   The average cost per mile is $5.

   $c(m) = 5(m - 1) + 10$

   $c(m) = 5m + 5$

6. **(B)**

   For the given graph, the horizontal asymptote is $y = 3$, y-intercept is $(0, 2)$ and It passes through $(-1, 1)$.

7. **(C)**

   Line k is perpendicular to the line l. The slope of line k is an opposite reciprocal of 4, which is $-\frac{1}{4}$. (C) $y = -\frac{1}{4}x + 5$

8. **(C)**

   x-intercept is $(-6, 0)$ and y-intercept is $(0, -8)$.

9. **(C)**

$F(t) = 60 \cdot (\frac{1}{2})^{0.02t} = 60 \cdot (\frac{1}{2})^{\frac{t}{50}}$

The decay period is 50 minutes.

10. **(A)**

x-intercept is $(600, 0)$ and y-intercept is $(0, 5)$.

$0.5c + 60p = 300$ satisfies the intercepts.

11. **(B)**

$g(x)$ is translated from $f(x)$ 2 units to the right.

When c is positive, $f(x - c)$ represents horizontal shift c units to the right.

$f(x + c)$ represents horizontal shift c units to the left.

12. **4**

$50 + 40x = 210$

$x = 4$

13. **(D)**

The total monthly revenue from each calculator sold is 120.

The total monthly revenue from selling x calculators is $120x$.

14. **(C)**

$a + b = 5,000$

$|a - b| \le 500 \rightarrow -500 \le a - b \le 500$

15. **130,131,132**

Let w be the number of minutes walking and s be the number of minutes swimming.

$s = 120$ minutes.

$1,600 \le 4w + 9(120) \le 1,608$

$130 \le w \le 132$

16. **(A)**

$\frac{1}{2} = \frac{1}{2} \ne 1$

The system has ZERO solution.

17. **$\frac{2}{3}$**

x-intercept is $(-\frac{12}{a}, 0)$ and y-intercept is $(0, -\frac{12}{b})$.

$-\frac{12}{a} = -6, a = 2$

$-\frac{12}{b} = -4, a = 3$

18. **24**

$1 - 0.76 = 0.24$

$0.24 \times 100 = 24\%$

19. **4.03**

$\frac{\Delta s}{\Delta d} = 800, \frac{3224}{\Delta d} = 800.$

$\Delta d = 4.03$

20. **729**

$3^{\frac{t}{20}} = 3^{\frac{2 \times 60}{20}} = 3^6 = 729$

21. **440**

$2(x+5) = 43 + 67 = 110$

$8(x+5) = 4 \times 2(x+5) = 440$

22. **(C)**

The growth factor is approximately 1.28.

23. **(B)**

Initial lux of illumination is 3,300. The decay factor is $1 - 0.02 = 0.98$

$f(t) = 3,300(0.98)^t$

Chapter 2

Quadratic Function and
Polynomial Function

2.1 Quadratic Function

> ### Quadratic functions
>
> 3 forms of quadratic functions are given below.
>
> 1. Standard form : $f(x) = ax^2 + bx + c, a \neq 0$
>
> 2. Vertex form : $f(x) = a(x - h)^2 + k$, where vertex is $(h, k) = (-\frac{b}{2a}, f(-\frac{b}{2a}))$
>
> 3. Factored form : $f(x) = a(x - p)(x - q)$, where x-intercepts (roots, zeros, and solutions) are $x = p, q$
>
> *x-coordinate of vertex in factored form is $h = \frac{p+q}{2}$

2.1.1   Example

$$y = a(x + 2)(x - 4)$$

In the given function above, the minimum value of y is -27, what is the value of a ?

2.1.2   Example

$$y = x^2 - 34x + k$$

In the function equation, k is a constant. The function has a minimum value at (x, y). What is the value of y ?

 (A) k

 (B) $-28 + k$

 (C) $867 + k$

 (D) $-289 + k$

Definition 5: Factoring, Quadratic formula, Discriminant, and Sum and Product of zeros

1. Factoring formula

   $a^2 - b^2 = (a - b)(a + b)$

   $a^3 - b^3 = (a - b)(a^2 + ab + b^2)$

   $a^3 + b^3 = (a + b)(a^2 - ab + b^2)$

   $a^2 + 2ab + b^2 = (a + b)^2$

   $a^2 - 2ab + b^2 = (a - b)^2$

   $a^3 + 3a^2b + 3ab^2 + b^3 = (a + b)^3$

   $a^3 - 3a^2b + 3ab^2 - b^3 = (a - b)^3$

2. Quadratic formula

   $f(x) = ax^2 + bx + c$

$$x = \frac{-b \pm \sqrt{b^2 - 4ac}}{2a}$$

3. Discriminant (for the number of solutions)

$$D = b^2 - 4ac$$

   - If $D > 0$, there are **2 real** zeros or roots.
   - If $D = 0$, there is **one real** zero or root.
   - If $D < 0$, there is **no real** zeros or roots, but there are 2 complex zeros or roots.

4. Sum and Product of 2 zeros

   $f(x) = ax^2 + bx + c = a(x - p)(x - q)$

$$p + q = -\frac{b}{a}$$

$$p * q = \frac{c}{a}$$

2.1.3 Example

$$\frac{1}{x^4} + \frac{2}{x^2 y^2} + \frac{1}{y^4}$$

Which of the following is equivalent to the given expression above?

(A) $\frac{1}{x^2} + \frac{1}{y^2}$

(B) $\left(\frac{1}{x} + \frac{1}{y}\right)^2$

(C) $\frac{1}{x^2} + \frac{\sqrt{2}}{xy} + \frac{1}{y^2}$

(D) $(x^{-2} + y^{-2})^2$

2.1.4 Example

$$2x^3 - 5x^2 - 6x + 15$$

Which of the following is a factor of the expression above?

(A) $2x + 5$

(B) $2x - 5$

(C) $x + 3$

(D) $x - 3$

2.1.5 Example

$$x^2 - 6x - 11 + k = 0$$

In the equation, k is a constant and it has exactly two equal roots. What is the value of k?

2.2 Complex Number

Definition 6: Complex number

$$i = \sqrt{-1}, \ i^2 = -1, \ i^3 = -i, \ i^4 = 1$$

1. The complex number is $a + bi$, where a is the real part and bi is the imaginary part.

2. The conjugate of a complex number is $a - bi$

3. $\mid a + bi \mid = \sqrt{a^2 + b^2}$

4. If $a + bi = c + di$, then $a = c$ and $b = d$

5. $(a + bi) + (c + di) = (a + c) + (b + d)i$

6. $(a + bi) - (c + di) = (a - c) + (b - d)i$

7. $(a + bi)(c + di) = (ac - bd) + (ad + bc)i$

8. $(a + bi)(a - bi) = a^2 + b^2$

9. $\frac{c+di}{a+bi} = \frac{(c+di)(a-bi)}{a^2+b^2}$

2.2.1 Example

Which of the following is equal to $\frac{-19-i}{10-9i}$?

    (A) $-\frac{199}{181} - i$

    (B) $-\frac{199}{181} + \frac{161}{181}i$

    (C) $-1 - i$

    (D) $-1 + i$

2.2.2 Example

$$\left(2 - \frac{3}{2}i\right)(4 + 3i) = (a + bi)$$

In the equation above, a and b are real numbers and $i = \sqrt{-1}$. What is the value of a?

2.2.3 Example

In the complex number system, what is the value of the expression $(-3 + 2i)(1 - i^3)$

(A) $-1 + 5i$

(B) $-1 - i$

(C) $-5 + 5i$

(D) $-5 - i$

2.3 Polynomial Function

1. **Equation of a polynomial function**

$$f(x) = a_n x^n + a_{n-1} x^{n-1} + \cdots + a_1 + a_0$$

n is the highest degree and it must be an positive integer.

a_n is the leading coefficient.

2. **Graph of a polynomial function**

There are two characteristic of polynomial functions; End-behavior and Multiplicity.

The end behavior of a polynomial is closely related to the end behavior of its leading term. Based on the leading coefficient and the highest degree, we can determine where ends go.

- **End behavior**

	$a_n > 0$	$a_n < 0$
$n =$ even	**Both Up** $\lim_{x \to -\infty} f(x) = +\infty$ $\lim_{x \to \infty} f(x) = +\infty$	**Both down** $\lim_{x \to -\infty} f(x) = -\infty$ $\lim_{x \to \infty} f(x) = -\infty$
$n =$ odd	**Down $\to$ Up** $\lim_{x \to -\infty} f(x) = -\infty$ $\lim_{x \to \infty} f(x) = +\infty$	**UP $\to$ Down** $\lim_{x \to -\infty} f(x) = +\infty$ $\lim_{x \to \infty} f(x) = -\infty$

- **Multiplicity**

The multiplicity is the number of times a factor occurs. A factor $(x - c)^k, k > 1$, yields a repeated zero $x = c$ of multiplicity k

(a) If $k = 1$, the graph **crosses** the x-axis at $x = c$

(b) If $k = odd \geq 3$, the graph has **a point of inflection** at $x = c$

(c) If $k = even$, the graph **bounces off** the x-axis at $x = c$

1. **Quotient-Remainder form**

$$\frac{P(x)}{(x-c)} = Q(x) + \frac{R}{(x-c)}$$

Example)

$$
\begin{array}{r}
6x - 2 \\
\hline
2x + 1 \,) \, 12x^2 + 2x + 4 \\
-(12x^2 + 6x) \\
\hline
-4x + 4 \\
-(-4x - 2) \\
\hline
6
\end{array}
$$

$$\frac{12x^2 + 2x + 4}{2x + 1} = 6x - 2 + \frac{6}{2x + 1}$$

2. **Remainder Theorem**

If the polynomial $P(x)$ is divided by $x - c$, then the remainder is $P(c)$

$$P(x) = (x - c)Q(x) + R$$

The remainder is $R = P(c)$, where $x = c$ is a solution of the divisor

3. **Factor Theorem**

$x = c$ is a zero of $P(x)$ if and only if $x - c$ is a factor of $P(x)$

$$P(x) = (x - c)Q(x)$$

$$P(c) = 0$$

1. Horizontal translation

   Let c be a positive constant.

   - $f(x - c)$: Horizontal shift c units to the right
   - $f(x + c)$: Horizontal shift c units to the left

2. Vertical translation

   Let d be a positive constant.

   - $f(x) + d$: Vertical shift d units upwards
   - $f(x) - d$: Vertical shift d units downwards

3. Reflection over axis

   - $f(-x)$: Reflection over the y-axis
   - $-f(x)$: Reflection over the x-axis

4. Dilation

   Let a and b be positive constants.

   - $f(ax)$: Horizontal dilation by a factor of $\frac{1}{a}$

     If $0 < a < 1$, It is horizontally stretched by a factor of $\frac{1}{a}$

     If $a > 1$, It is horizontally compressed by a factor of $\frac{1}{a}$

   - $bf(x)$: Vertical dilation by a factor of b

     If $0 < b < 1$, It is vertically compressed by a factor of b

     If $b > 1$, It is vertically stretched by a factor of b

2.3.1   Example

$$(x - 8)^2 + (y + 7)^2 = 9$$

The graph in the xy-plane of the equation above is a circle. If the circle is translated upward a units such that the circle is tangent to the x-axis, the equation becomes $(x - 8)^2 + (y + 7 - a)^2 = 9$. What is the value of a ?

2.3.2 Example

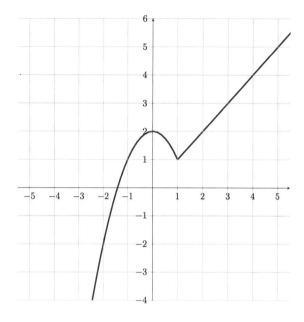

The complete graph of the function f is shown in the xy-plane above. What is the y intercept of the graph of $y = f(x - 2)$?

(A) -2

(B) 0

(C) 2

(D) 4

2.3.3 Example

If the polynomial $p(x) = x^3 + 3x^2 + kx - 2$ is divisible by $x + 2$, what is the value of k?

2.3.4 Example

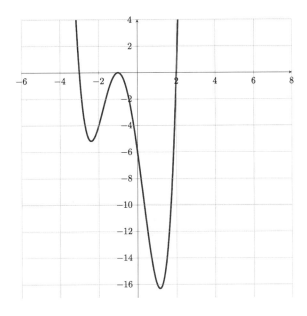

The graph of the function f is shown in the xy-plane above, where $y = f(x)$. Which of the following functions could define f?

(A) $f(x) = (x + 4)(x + 1)^2(x - 2)$

(B) $f(x) = (x + 4)^2(x + 1)(x - 2)$

(C) $f(x) = (x + 3)(x + 1)^2(x - 2)$

(D) $f(x) = (x + 3)^2(x + 1)(x - 2)$

2.3.5 Example

For a polynomial $t(x)$, the value of $t(0)$ is 2. Which of the following must be true about $t(x)$?

(A) $x + 2$ is a factor of $t(x)$.

(B) x is a factor of $t(x)$

(C) $x - 2$ is a factor of $t(x)$

(D) The remainder when $t(x)$ is divided by x is 2

2.1.1 Example

The x-intercepts are $x = -2, 4$. The x coordinate of the vertex is $\frac{-2+4}{2} = 1$ Then, the y coordinate of the vertex is $f(1) = -27$.

$-27 = a(3)(-3)$

Answer: a $= 3$

2.1.2 Example

The x coordinate of the vertex is $\frac{34}{2} = 17$. Then, the y coordinate of the vertex is $f(17) = 17^2 - 34(17) + k = -289 + k$.

Answer: (D)

2.1.3 Example

$$\frac{1}{x^4} + \frac{2}{x^2 y^2} + \frac{1}{y^4} = (\frac{1}{x^2})^2 + 2 \cdot \frac{1}{x^2} \cdot \frac{1}{y^2} + (\frac{1}{y^2})^2 = (x^-2 + y^-2)^2$$

Answer: (D)

2.1.4 Example

$$2x^3 - 5x^2 - 6x + 15 = (2x^3 - 5x^2) + (-6x + 15) = x^2(2x - 5) - 3(2x - 5) = (x^2 - 3)(2x - 5)$$

Answer: (B)

2.1.5 Example

$$D = b^2 - 4ac = (-6)^2 - 4(-11 + k) = 36 + 44 - 4k = 80 - 4k = 0$$

$$k = 20$$

Answer: k $= 20$

2.2.1 Example

$$\frac{-19 - i}{10 - 9i} = \frac{(-19 - i)(10 + 9i)}{(10 - 9i)(10 + 9i)} = \frac{-181 - 181i}{181} = -1 - i$$

Answer: (C)

2.2.2 Example

$$(2 - \frac{3}{2}i)(4 + 3i) = 8 + 6i - 6i + \frac{9}{2} = \frac{25}{2}$$

Answer: $\frac{25}{2} = 12.5$

2.2.3 Example

$i^3 = -i$, $(-3 + 2i)(1 - i^3) = -3 - 3i + 2i - 2 = -5 - i$

Answer: (D)

Explanation

2.3.1 Example

The center of the given circle is $(8, -7)$ and the radius is 3. The closet point on the circle to the x is $(3, -4)$. The circle is shifted 4 units up, the circle is tangent to the x-axis.

Answer: a = 4

2.3.2 Example

$y = f(x - 2)$ represents the horizontal shift 2 units to the right.

$(-2, -2) \rightarrow (0, -2)$

Answer: (A)

2.3.3 Example

By the factor theorem, $p(-2) = -8 + 12 - 2k - 2 = 0$

$k = 1$

Answer: k = 1

2.3.4 Example

The graph of the function f has zeros at $x = -3, -1$, and 2.

$(x + 3)$ has multiplicity 1.

$(x + 1)$ has multiplicity 2.

$(x - 2)$ has multiplicity 1.

Answer: (C)

2.3.5 Example

By the remainder theorem, the remainder is 2 when $t(x)$ is divided by x.

Answer: (D)

2.4 Chapter 2 from A to Z Problems

1.

$$h(t) = -5t^2 + 18t + 8$$

The function h above can be used to model the height above the ground, in meters, of an object t seconds after the object is launched straight up from the initial height of 8 meters. According to the model, what is the maximum height the object can reach from the ground?

2.

$$3(x + 5)^2 + k = 4$$

If k is a constant,the quadratic equation given above have exactly two solutions. Which of the following could be a value of k?

(A) -3

(B) 4

(C) 5

(D) 6

3. A solution of the quadratic function $f(x) = ax^2 + 13x - 6$ is equal to -3. Which of the following is a factor of f?

(A) $5x + 2$

(B) $5x - 2$

(C) $x - 3$

(D) $x + 5$

4. In the xy-plane, the graph of the equation $y = 10x - 18$ intersects the graph of the equation $y = x^2$ at two points. What is the sum of the x-coordinates of the two points?

(A) -18

(B) -10

(C) 10

(D) 18

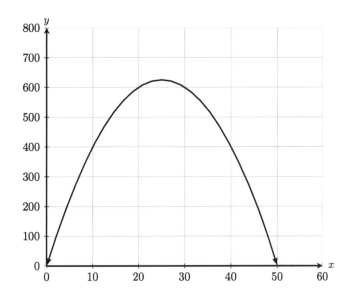

5. The graph shown models the revenue y, in thousands of dollars, of ticket sales and the number of attendance x, in thousands, at a small theater. Which equation represents this model?

(A) $y = -x^2 + 625$

(B) $y = -(x - 50)^2$

(C) $y = 625 - (x - 25)^2$

(D) $y = 625 + (x - 25)^2$

6. Which of the following complex numbers is equivalent to $\frac{5+i}{5-i}$?

(A) $\frac{12}{13} + \frac{5}{13}i$

(B) $\frac{12}{13} - \frac{5}{13}i$

(C) $\frac{24}{25} + \frac{2}{5}i$

(D) $\frac{24}{25} - \frac{2}{25}i$

7.

$$(3x - 4y) + (6x + 2y)i = 5i$$

In the equation above, x and y are real numbers and $i = \sqrt{-1}$. What is the value of x ?

8.

$$\frac{-16}{x^2 - 16} = \frac{2}{x + 4} - \frac{1}{x - 4}$$

Which statement describes the solution to the equation above?

(A) 4 is the only solution

(B) -4 is the only solution

(C) -4 and 4 are both solutions

(D) No value of x satisfies the equation

9. A ball is thrown upward from a height of 3 feet above the ground. Assuming no air resistance, the function $h(t)$ defined by $h(t) = -16t^2 + 36t + 3$ models the balls height $h(t)$, in feet, above the ground t seconds after it is thrown. According to the model, at what time, in seconds, does the ball hit the maximum height?

(A) 1.125

(B) 2

(C) 11

(D) 23.25

10. The graph of a certain parabola in the xy-plane intersects the y-axis once and the x-axis twice. Which of the following forms of the parabola's equation gives the x-value of the x-intercepts as a constant?

(A) $f(x) = (2x - 3)(5x + 7)$

(B) $f(x) = 10(x - \frac{3}{2})(x + \frac{7}{5})$

(C) $f(x) = 2(x - \frac{3}{2})(5x + 7)$

(D) $f(x) = 5(2x - 3)(x + \frac{7}{5})$

11.

$$243x^{36} - 75y^{16}$$

Which of the following is equivalent to the above expression?

(A) $3(9x^6 - 5y^4)(9x^6 + 5y^4)$

(B) $3(81x^6 - 25y^4)(81x^6 + 25y^4)$

(C) $3(9x^{18} - 5y^8)(9x^{18} + 5y^8)$

(D) $3(81x^{18} - 25y^8)(81x^{18} + 25y^8)$

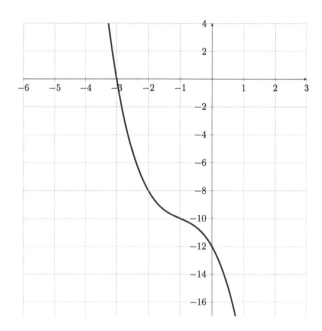

12. The graph of the function f is shown in the xy-plane above, where $y = f(x)$. Which of the following functions could define f?

(A) $f(x) = (x + 3)(x^2 - 4)$

(B) $f(x) = (x - 3)(x^2 + 4)$

(C) $f(x) = -(x + 3)(x^2 - 4)$

(D) $f(x) = -(x + 3)(x^2 + 4)$

13.

$$\frac{k^2 - 1}{4k - 28} + \frac{48}{28 - 4k}$$

Which of the following is equivalent to the given expression, where k is not equal to -7?

(A) $\frac{-k^2 - 49}{4k + 28}$

(B) $\frac{k^2 + 47}{4k + 28}$

(C) $\frac{k + 7}{4}$

(D) $k + \frac{7}{4}$

14.

$$p(x) = (15x^2 - 42)(x - k) - 168$$

In the polynomial $p(x)$ defined above, k is a constant. If x is a factor of $p(x)$, what is the value of k?

15. Which of the following expressions is equivalent to $\frac{x^2 - 2x - 4}{x + 3}$?

(A) $x - 5 - \frac{19}{x+3}$

(B) $x - 5 + \frac{11}{x+3}$

(C) $x + 1 - \frac{7}{x+3}$

(D) $x + 1 - \frac{1}{x+3}$

16.

$$(x - 1)(x + 2)(x - 5) > 0$$

In the inequality above, which of the following could be a solution ?

(A) -3

(B) 1

(C) 4

(D) 8

17. The function is defined by $f(x) = (x + 2)^2$. If $f(x + a) = x^2 - 10x + 25$, where a is a constant, what is the value of a?

(A) -7

(B) -5

(C) 5

(D) 7

18. Twice the length of a rectangle is three times its width. If the width of the rectangle is x feet, which of the following function $A(x)$ could represent the area, in square feet, of the rectangle?

(A) $A(x) = \frac{3}{2}x^2$

(B) $A(x) = \frac{2}{3}x^2$

(C) $A(x) = \frac{9}{4}x^2$

(D) $A(x) = \frac{4}{9}x^2$

19.

$$7x^3 + bx^2 - 21x - 3b$$

In the polynomial above, b is constant. Which of the following is a factor of the polynomial?

(A) $x + 3$

(B) $x - 3$

(C) $7x + b$

(D) $7x^3 + b$

20.

$$2x^2 + bx + 8 = 0$$

In the equation above, b is a constant. If the equation has No real solution, what could be the value of b ?

(A) -10

(B) -8

(C) 5

(D) 10

21.

$$x^2 + bx - 36 = 0$$

In the equation, b is a positive integer constant. Which value could be a solution to the equation?

(A) 1

(B) 9

(C) 12

(D) 18

22.

$$f(n) = 158n^2 - 771n + 10,268$$

The number of farms, in millions, in the United States n years after 1949 can be modeled by the function f above, for $0 \leq n \leq 30$. The constant term 10,268 in the function is an estimate for which of the following?

(A) The number of farms, in millions, in 1949

(B) The number of farms, in millions, in 1979

(C) The increase in the number of farms, in millions each year

(D) The maximum number of farms in a single year from 1949 through 1979

23.

$$x^2 - 8x = 5$$

The equation above has solutions $x = 4 + \sqrt{n}$ and $x = 4 - \sqrt{n}$, where n is a positive integer. What is the value of n?

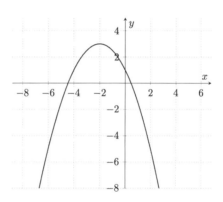

24. The graph of $y = r(x)$ is shown in the xy-plane. If $a, h,$ and k are positive constants, which of following could define the function r?

(A) $r(x) = -a(x + h)^2 + k$

(B) $r(x) = -a(x - h)^2 + k$

(C) $r(x) = a(x + h)^2 + k$

(D) $r(x) = a(x - h)^2 + k$

2.4 ANSWERS							
Number	Answer	Number	Answer	Number	Answer	Number	Answer
1	24.2	7	$\frac{2}{3}$	13	C	19	C
2	A	8	D	14	4	20	C
3	B	9	A	15	B	21	A
4	C	10	B	16	D	22	A
5	C	11	C	17	A	23	21
6	A	12	D	18	A	24	A

2.5 Explanation

1. **24.2**

   $h = -\frac{b}{2a} = \frac{-(-18)}{10} = 1.8$

   $k = h(1.8) = 24.2$

2. **(A)**

   $(x+5)^2 = \frac{4-k}{3} > 0$

   $4 - k > 0$

   $k < 4$

3. **(B)**

   $f(-3) = 9a - 39 - 6 = 0, a = 5$

   $5x^2 + 13x - 6 = (5x - 2)(x + 3)$

4. **(C)**

   $x^2 = 10x - 8, x^2 - 10x + 18 = 0$ The sum of zeros is $-\frac{b}{a} = 10$

5. **(C)**

   The vertex is $(25, 625)$. It opens downward, so the leading coefficient must be negative.

6. **(A)**

   $\frac{5+i}{5-i} = \frac{(5+i)^2}{(5-i)(5+i)} = \frac{24+10i}{26} = \frac{12}{13} + \frac{5i}{13}$

7. **$\frac{2}{3}$**

   $3x - 4y = 0, 6x + 2y = 5$

   $x = \frac{2}{3}, y = \frac{1}{2}$

8. **(D)**

   Undefined x-values are $x = 4, -4$

   $-16 = 2(x - 4) - (x + 4)$

   $x = -4$

   No solution

9. **(A)**

$h = -\frac{b}{2a} = \frac{36}{2(16)} = \frac{18}{16} = \frac{9}{8} = 1.125$

10. **(B)**

Factored form is $a(x-p)(x-q)$.

11. **(C)**

$3(81x^{36} - 25y^{16}) = 3((9x^{18})^2 - (5y^8)^2) = 3(9x^{18} - 5y^8)(9x^{18} + 5y^8)$

12. **(D)**

As x goes to negative infinity, y approaches to positive infinity and as x goes to positive infinity, y approaches to negative infinity.

The leading coefficient is negative and it has one zero.

(D) $x = -3, \pm 2i$

13. **(C)**

$\frac{k^2-1}{4(k-7)} + \frac{48}{-4(k-7)} = \frac{k^2-49}{4(k-7)} = \frac{(k+7)(k-7)}{4(k-7)} = \frac{k+7}{4}$

14. **4**

$p(0) = -42(-k) - 168 = 0$

$42k = 168$

$k = 4$

15. **(B)**

$$
\begin{array}{r}
x - 5 \\
x+3 \enclose{longdiv}{x^2 - 2x - 4} \\
\underline{-(x^2 + 3x)} \\
-5x - 4 \\
\underline{-(-5x - 15)} \\
11
\end{array}
$$

$$\frac{x^2 - 2x - 4}{x + 3} = x - 5 + \frac{11}{x + 3}$$

16. **(D)**

When $-2 < x < 1$ or $x > 5$, y-values are positive.

17. **(A)**

$f(x + a) = (x - 5)^2$

$f(x) = (x + 2)^2 \rightarrow f(x + a)^2 = (x - 5)^2$ $f(x)$ is translated 7 units to the right.

$a = -7$

18. **(A)**

$2l = 3w = 3x$ $l = \frac{3}{2}x$ $A(x) = \frac{3}{2}x^2$

19. **(C)**

$7x^3 + bx^2 - 21x - 3b = x^2(7x + b) - 3(7x + b) = (x^2 - 3)(7x + b)$

20. **(C)**

$b^2 - 4(2)(8) < 0$

$b^2 - 64 < 0$

$b^2 < 64$

21. **(A)**

Since b is positive, possible zeros are $x = 1, -36$, $x = 2, -18$, $x = 3, -12$, or $x = 4, -9$

22. **(A)**

10,268 represents y-intercept and it means the number of farms, in millions when 0 years after 1949.

In 1949, the number of farms, in millions is 10,268.

23. **21**

$x^2 - 8x - 5 = 0$

$x = \frac{8 \pm \sqrt{84}}{2} = \frac{8 \pm 2\sqrt{21}}{2} = 4 \pm \sqrt{21}$

24. **(A)**

The equation of the given graph is $y = -\frac{1}{2}(x + 2)^2 + 3$

Chapter **3**

Radical Expressions and Equations

3.1 Exponents and Radicals

Rules	Equivalent form	Example
a^0	1	1. $-3(a)^0$
a^{-x}	$\frac{1}{a^x}$	2. $\left(\frac{2}{3}\right)^{-2}$
$a^x \cdot a^y$	a^{x+y}	3. $x^{5/6} \cdot x^{6/5}$
$\frac{a^x}{a^y}$	a^{x-y}	4. $\frac{x^{5/6}}{x^{6/5}}$
$\sqrt[n]{a}$	$a^{\frac{1}{n}}$	5. $\frac{1}{\sqrt{3}}$
$\sqrt[n]{a^m}$	$a^{\frac{m}{n}}$	6. $\frac{1}{\sqrt{3^5}}$
$(a^x)^y$	a^{xy}	7. $(-2x^{-3}y^{-1})^3$
$\sqrt[m]{\sqrt[n]{a}}$	$a^{\frac{1}{mn}}$	8. $\sqrt[3]{\sqrt{64}}$
$\sqrt[n]{ab}$	$a^{\frac{1}{n}}b^{\frac{1}{n}}$	9. $2^{\frac{2}{3}} \cdot 3^{\frac{4}{3}}$
$\sqrt[n]{\frac{a}{b}}$	$\frac{a^{\frac{1}{n}}}{b^{\frac{1}{n}}}$	10. $\frac{\sqrt[3]{4}}{\sqrt[3]{108}}$
$\sqrt[n]{a^n}, n = odd$	a	11. $\sqrt[3]{(x-3)^3}$
$\sqrt[n]{a^n}, n = even$	$\lvert a \rvert$	12. $\sqrt{(x-3)^2}$

3.2 Radical Equations

3.2.1 Example

$$\sqrt{3x + 4} = 2x + 1$$

what is the solution to the above equation?

3.2.2 Example

$$\sqrt[3]{x^2} \cdot \sqrt{x^3} = x^k$$

What value of k satisfies the above equation?

3.2.3 Example

$$\sqrt{(5 - x)^2} = 32$$

What is the sum of all solutions to the above equation?

(A) -27

(B) -10

(C) 10

(D) 37

3.1 Example

1. -3
2. $\frac{9}{4}$
3. $x^{\frac{61}{30}}$
4. $\frac{1}{x^{\frac{11}{30}}}$
5. $\frac{1}{3^{\frac{1}{2}}}$
6. $\frac{1}{3^{\frac{5}{2}}}$
7. $-\frac{8}{x^9 y^3}$
8. 2
9. $3\sqrt[3]{12}$
10. $\frac{1}{3}$
11. $x - 3$
12. $|x - 3|$

3.2.1 Example

Limited domain : $x \geq -\frac{1}{2}$

$$(\sqrt{3x+4})^2 = (2x+1)^2$$

$$4x^2 + x - 3 = (4x-3)(x+1) = 0$$

$x = \frac{3}{4}, -1$ but $x = -1$ cannot be a root.

Answer: $\frac{3}{4}$

3.2.2 Example

$$\sqrt[3]{x^2} \cdot \sqrt{x^3} = x^{\frac{2}{3}} \cdot x^{\frac{3}{2}} = x^{\frac{2}{3}+\frac{3}{2}} = x^{\frac{13}{6}}$$

$k = \frac{13}{6}$

Answer: $\frac{13}{6}$

3.2.3 Example

$$\sqrt{(5-x)^2} = 32$$

$$|5 - x| = 32$$

$$5 - x = \pm 32$$

$$x = 37, -27$$

The sum of solutions is 10.

Answer: (C)

3.3 Chapter 3 from A to Z Problems

1.

$$\sqrt[3]{p^2} = \sqrt[5]{q}$$

$$p^{2x} = q^6$$

In the equation above, p and q are constants, $p > 1$, and $q > 1$. What is the value of x?

2. Which of the following expression is equivalent to $x^{\frac{3}{2}}\left(\sqrt{x} - \frac{1}{\sqrt{x}}\right)$?

(A) $x^2 - 1$
(B) $\sqrt[4]{x^3} - \frac{1}{\sqrt[4]{x^3}}$
(C) $\sqrt{x^7} - \frac{1}{\sqrt{x}}$
(D) $x^2 - x$

3.

$$\sqrt{x + a} = x - 9$$

If $a = 33$, what is the solution set of the equation above?
(A) $\{3, 16\}$
(B) $\{3\}$
(C) $\{9\}$
(D) $\{16\}$

4.

$$4^{\frac{1}{5}} \cdot 3^{\frac{2}{5}} = y^{\frac{1}{5}}$$

What value of y satisfies the above equation?

5.

$$\sqrt{x^2 + 3} = x - 5$$

Which of the following could be a solution for the equation above?
(A) 0
(B) 2.2
(C) 5
(D) No value of x satisfies the equation

6.

$$\sqrt[3]{27x^3y^5 - 81x^4y^3}$$

Which of the following is an equivalent form of the above equation?

(A) $3xy\sqrt[3]{y^2 - 3x}$
(B) $3xy(\sqrt[3]{y^2} - \sqrt[3]{3x})$
(C) $27xy\sqrt[3]{y^2 - 3x}$
(D) $27xy(\sqrt[3]{y^2} - \sqrt[3]{3x})$

7.

$$\sqrt{4x} + \sqrt{9x} = 4$$

What is the solution to the above equation?

8.

$$\sqrt{(2x+1)^2} = 2x + 1$$

Which of the following values of x is NOT a solution to the equation above?

(A) -2

(B) $-\frac{1}{2}$

(C) 0

(D) 2

9. The function p is defined by $p(x) = 2^x$. If $p(k^2) \cdot p(2) = 8$, what is the value of k?

(A) -2

(B) -1

(C) $\sqrt{6}$

(D) $\frac{\sqrt{6}}{2}$

10. How many values of x satisfy the equation $\sqrt{x+6} = x$

    (A) None

    (B) One

    (C) Two

    (D) More than two

11. Which of the following expression is equivalent to $(\sqrt{z} + \sqrt{z^3})^2$?

    (A) z^4

    (B) $z + z^3$

    (C) $z^{\frac{5}{2}} + z^{\frac{7}{2}}$

    (D) $z + 2z^2 + z^3$

12.

$$Q = \sqrt{\frac{2DS}{H}}$$

The formula above is used to estimate the optimal number of product units, Q, for a company to order given the demand quantity, D; the order cost per item, S: and the holding cost per item, H. Which of the following correctly express the holding cost per item in terms of other variables?

    (A) $H = \sqrt{\frac{2DS}{Q}}$

    (B) $H = \frac{\sqrt{2DS}}{Q}$

    (C) $H = \frac{2DS}{Q^2}$

    (D) $H = \frac{Q^2}{2DS}$

13.

$$\sqrt{(x+y)^3} = 27$$

What is the value of $(x+y)^2$?

14. Which of the following expression is equivalent to $(\frac{\sqrt{q}}{2} + \frac{\sqrt{r}}{2})^2$?

 (A) $\frac{q}{4} + \frac{r}{4}$

 (B) $\frac{q}{4} + \frac{\sqrt{qr}}{2} + \frac{r}{4}$

 (C) $\frac{1}{4}(q + qr + r)$

 (D) $\frac{1}{4}(q + \sqrt{qr} + r)$

3.3 ANSWERS					
Number	Answer	Number	Answer	Number	Answer
1	10	7	$\frac{16}{25}$	13	81
2	D	8	A	14	B
3	D	9	B	15	
4	36	10	B	16	
5	D	11	D	17	
6	A	12	C	18	

3.4 Explanation

1. **10**

   $(p^{\frac{2}{3}})^{30} = (q^{\frac{1}{5}})^{30}$

   $p^{20} = p^{2x}$

   $x = 10$

2. **(D)**

   $x^{\frac{3}{2}}(x^{\frac{1}{2}} - x^{-\frac{1}{2}}) = x^2 - x$

3. **(D)**

   $\sqrt{x + 33} = x - 9$

   Limited domain is $x \geq 9$.

   $x + 33 = (x - 9)^2$

   $x^2 - 18x + 81 - x - 33 = 0$

   $x^2 - 19x + 48 = 0$

   $x = 3, 16 \ x \neq 3, x = 16$

4. **36**

   $(4 \cdot 3^2)^{\frac{1}{5}} = 36^{\frac{1}{5}}$

5. **(D)**

   Limited domain is $x \geq 5$. $(\sqrt{x^2 + 3})^2 = (x - 5)^2$

   $x^2 + 3 = (x - 5)^2$

   $x^2 - 10x + 25 - x^2 - 3 = 0$

   $10x = 22$

   $x = 2.2$ Since the value is less than 5, it cannot be a solution.

6. **(A)**

   $\sqrt[3]{27x^3y^5 - 81x^4y^3} = 3xy\sqrt[3]{y^2 - 3x}$

7. $\frac{16}{25}$

$$2\sqrt{x} + 3\sqrt{x} = 4$$
$$5\sqrt{x} = 4$$
$$\sqrt{x} = \frac{4}{5}$$
$$x = \frac{16}{25}$$

8. **(A)**

$$\sqrt{(2x+1)^2} = 2x + 1$$
$$|2x+1| = 2x + 1$$
$$x \geq -\frac{1}{2}$$

9. **(B)**

$$2^{k^2} \cdot 2^2 = 2^{k^2+2} = 2^3$$
$$k^2 = 1$$
$$k = \pm 1$$

10. **(B)**

Limited domain is $x \geq 0$. $x + 6 = x^2$

$$x^2 - x - 6 = (x-3)(x+2) = 0$$
$$x = -2, 3$$

But $x = -2$ is an extraneous root, so it must be rejected.

11. **(D)**

$$(\sqrt{z} + \sqrt{z^3})^2 = z + 2\sqrt{z}\sqrt{z^3} + z^3 = z + 2z^2 + z^3$$

12. **(C)**

$$Q^2 = \frac{2DS}{H}$$
$$H = \frac{2DS}{Q^2}$$

13. **81**

$$(x+y)^3 = 27^2$$
$$x + y = (27)^{\frac{2}{3}}$$
$$x + y = 9$$
$$(x+y)^2 = 81$$

14. **(B)**

$$\left(\frac{\sqrt{q}}{2} + \frac{\sqrt{r}}{2}\right)^2 = \frac{q}{4} + \frac{2\sqrt{qr}}{4} + \frac{r}{4}$$

Chapter **4**

Percent

4.1 Percent

4 types of percent problems

1. $p\%$ **of** x

$$p\% \ of \ x = \frac{p}{100} \times x$$

2. $p\%$ **more than** x or x **is increased by** $p\%$

$$p\% \ more \ than \ x = x \times (1 + \frac{p}{100})$$

3. $p\%$ **less than** x or x **is decreased by** $p\%$

$$p\% \ less \ than \ x = x \times (1 - \frac{p}{100})$$

4. **Percent change**

Percent Increase/Decrease=

$$\frac{New - Original}{Original} \times 100$$

4.1.1   Example

Renata opens her own bakery and sells cupcakes. In November, she sold 2.3 times as many cupcakes as she had in May. What was the percent increase in her number of cupcakes for this time period?

(A) 2.3%

(B) 23.0%

(C) 123.0%

(D) 130.0%

4.1.2   Example

K2 is the second-highest mountain in the world, with a total height of 8611 meters. The highest mountain in the world, Mount Everest, is approximately 1.028 times the height of mountain K2. The height of Mount Everest is what percent greater than the height of K2?

   (A) 2.80%
   (B) 28.0%
   (C) 102.8%
   (D) 202.8%

4.1.3   Example

a is 8% of b and b is 150% greater than c. If c is 25, what is the value of a?

4.1.4   Example

The number of occupational therapy jobs in the United States in 2012 has grown by 29% 10 years later. If occupational therapy jobs in 2022 were 145,800, how many occupational therapy jobs in 2012 to the nearest hundreds?

    (A) 103,500

    (B) 113,000

    (C) 141,600

    (D) 141,700

4.1.5   Example

In a particular store, the number of laptops sold the week of Black Friday was 685. The number of laptops sold the following week was 500. Laptop sales the week following Black Friday were what percent less than laptop sales the week of Black Friday to the nearest percent?

    (A) 17%

    (B) 27%

    (C) 37%

    (D) 47%

4.2 Rate

Rate

$$Distance = Speed \times Time$$

$$Time = \frac{Distance}{Speed}$$

$$Speed = \frac{Distance}{Time}$$

4.2.1 Example

A group of friends started a road trip in Georgia, a total distance of 112 miles. During the first x miles of driving in Georgia, they traveled an average of 8 miles per hour due to slow traffic. They drove an average of 55 miles per hour for the rest of the trip. The total drive time in Georgia was 2.25 hours. To the nearest <u>minute</u>, how many minutes did the group of friends drive at a speed of 8 miles per hour?

4.2.2 Example

In each revolution, the car travels a distance equal to the circumference of one of its tires. If the radius of a tire is 0.30 meters and the car travels at a rate of 70 revolutions per minute, what is the approximate speed of the car to the nearest <u>kilometer per hour</u>? (1 kilometer=1000 meters)

4.2.3 Example

A certain car can hold 22 gallons of fuel and travels at a constant speed of 40 miles per hour. At this speed, the car gets 25 miles per gallon. How many gallons remain when the car has traveled at this speed for 1 hour and 30 minutes?

4.2.4 Example

Pioneer 11 is a robotic space probe launched by NASA to observe the asteroid belt. If *Pioneer 11* is currently traveling at a rate of 11.378 kilometers per second, how fast is it traveling, in miles per hour? (1 miles is approximately 1.61 kilometers)

(A) 424.02

(B) 1,099.10

(C) 25,441.49

(D) 65,946.89

4.3 Unit Conversion-Squared Units and Cubic Units

> **Unit conversion**
>
> A length is measured in units, area is measured in square units, and volume is measured in cubic units.
>
> 1.
> $$1\text{yd} = 3 \text{ ft}$$
> $$1\text{yd}^2 = 9 \text{ ft}^2$$
> $$1\text{yd}^3 = 27 \text{ ft}^3$$
>
> 2.
> $$1\text{m} = 100 \text{ cm}$$
> $$1\text{m}^2 = 10,000 \text{ cm}^2$$
> $$1\text{m}^3 = 1,000,000 \text{ cm}^3$$

4.3.1 Example

The area of a parking lot in a certain business district is 3 square miles. What is the area, in square yards, of the parking lot? (1 mile= 1,760 yards)

    (A) 5,280
    (B) 15,840
    (C) 3,097,600
    (D) 9,292,800

4.3.2 Example

The estimated area of forest cover in the Brazilian Amazon, in square kilometers, in 1977 is 3,960,000 km^2. How many square megameters of forest cover was in 1977? (1000 kilometers= 1 megameter)

4.1.1 Example

Let N=the number of cupcakes sold in November.

Let M=the number of cupcakes sold in May.

$N = (2.3)M = (1 + 1.3)M$

Percent increase is 130%.

Answer: (D)

4.1.2 Example

The height of Mount Everest is $(1.028) \cdot 8611$

Percent greater is $0.028 \cdot 100 = 2.8\%$

Answer: (A)

4.1.3 Example

$a = 0.08b, b = 2.5c = 62.5, a = 0.08(62.5) = 5$

Answer: 5

4.1.4 Example

Let $x = $ the number of occupational therapy jobs in 2012.

$145,800 = (1.29) \cdot x$

$x = 113,000$

Answer: (B)

4.1.5 Example

Percent less $= \frac{500-685}{685} \times 100 = -27\%$

Answer: (B)

4.2.1 Example

$$\frac{x}{8} + \frac{112 - x}{55} = 2.25$$

$$x = 2\text{miles}$$

$$T = \frac{2}{8}\text{hours} = 15\text{minutes}$$

Answer: 15

4.2.2 Example

$$\frac{70 \times 2\pi \times 0.3 \times 60}{1000} = 7.9168 \approx 8$$

Answer: 8

4.2.3 Example

Distance $= 40 \times 1.5 = 60\text{miles}$ The remaining gallons is $22 - \frac{60}{25} = 22 - 2.4 = 19.6$

Answer: 19.6

Explanation

4.2.4 Example

$$11.378 \text{kilometerspersecond} = \frac{11.378}{1.61} \text{milespersecond}$$

$$\frac{11.378 \times 3600}{1.61} \text{milesperhour} = 25,441.49$$

Answer: (C)

4.3.1 Example

$$3 \text{miles}^2 = 3 \times 1760^2 \text{yd}^2 = 9,292,800$$

Answer: (D)

4.3.2 Example

$$1 \text{megameter}^2 = 1,000,000 \text{km}^2$$

$$\frac{3960000}{1000000} = 3.96$$

Answer: 3.96

4.4 Chapter 4 from A to Z Problems

1. In a particular soft drink manufacturing company, there are three operating machines A, B, and C. Every day, Machine A produces 30% more soft drinks than machine B, and machine B produces 15% less soft drinks than machine C. If machine A produces x soft drinks on a particular day, what is the total number of soft drinks produced on that day in the company in terms of x?

   (A) $x + 1.3x + 0.85x$
   (B) $x + 1.3x + 1.105x$
   (C) $x + \frac{x}{0.3} + \frac{x}{0.255}$
   (D) $x + \frac{x}{1.3} + \frac{x}{1.105}$

2. In 2020, the tourists of country A and country B were equal. From 2000 to 2020, the tourists of country A increased by 20% and the tourists of country B decreased by 10%. If the number of tourists in country A was 120,000 in 2000, how many more tourists visited country B in 2000?
   (A) 30,000
   (B) 40,000
   (C) 60,000
   (D) 120,000

3. Alan would like to reduce his yearly expenditure on gasoline. He spent $15,500 in gasoline last year and he is planning to reduce his expenditure by $2,325.00. By what percent would he save the cost this year ?

4. The heat capacity of a substance is the amount of energy, in joules (J), required to raise the temperature of 1 gram(g) of the substance by 1 degree Celsius ($°C$). The heat capacity of Oxygen is approximately 0.92 joules/gram degree Celsius ($\frac{J}{g \cdot °C}$). Approximately how much energy, in joules, is required to raise the temperature of 1.0 g of Oxygen from $-183°C$ to 22 $°C$, to the nearest whole number?

5. The elk herd in Yellowstone National Park is the sum of its males and females. The female elk is approximately 480, which is 60 % of the total elk in the park. Based on this approximation, how many males would be in the herd?

6. $p\%$ less than x is n. Which expression represents x in terms of p?

   (A) $\frac{100}{p}n$

   (B) $\frac{(100-p)n}{100}$

   (C) $\frac{100n}{100-p}$

   (D) $\frac{100n}{100+p}$

7. In a certain library, the number of books are increased by 3% every year from 2002 to 2014. There are x books in the library in 2002. Which expression represents the number of books in the library in 2014 in terms of x ?

(A) $x(1.03)$

(B) $x(1.03)^{12}$

(C) $x(0.03)$

(D) $x(1 + \frac{0.03}{12})^{12}$

8. A local restaurant gives a 20% discount on all their meals. Selena paid \$140.75, including 15% tips on the discounted price. What was the original price of the meals?

(A) \$129.490

(B) \$152.989

(C) \$202.328

(D) \$291.353

9. Kristin should pay the state income tax at the rate of 5% of the first \$28,000 of annual income plus 7% of any amount above \$28,000. If she will get paid \$70,000 this year, how much would she expect to pay for the tax?

10. The table summarizes data about a sample of 80 adult female polar bears and 30 adult male polar bears by their age.

Age(years)	Number of adult females	Number of adult males
5-9	45%	51%
10-14	32.5%	23%
15-19	10%	15%
20 or older	12.5%	11%
Total	100%	100%

What percentage of adult polar bears in the sample were 5 to 9 years old?

(A) 11.4%

(B) 46.6%

(C) 48%

(D) 96%

11. A pallet of sod costs $130, and covers approximately 450 square feet. A front lawn is represented by a square with a side length 75 ft. If the sod covers it, how much does the sod cost?

12. According to Ohm's law, the current of I through a conductor between two points is directly proportional to the voltage of E across the two points. If a current of $4I$ is applied to a certain electrical circuit, by what volt does the voltage change in terms of E?

(A) $E + 4$

(B) $E - 4$

(C) $\frac{E}{4}$

(D) $4E$

13. The ratio of the number of patents applied for in North America to the number of patents worldwide was 1:5 in 1990. There were a total of 200 patents, in thousands, applied for in North America. How many patents were applied for outside of North America in 1990?

(A) 800

(B) 1,000

(C) 800,000

(D) 1,000,000

14. According to a recipe for a chocolate chip cookie, 30 cookies calls for $2\frac{1}{4}$ cups of flour. If a cook plans to make fifteen dozen cookies, how many cups of flour should be used?

15. A car salesperson receives a monthly income $3000 and 5% of the amount of sales as commission. His goal is to get paid $12,500 this month, including the commission. Which of the following should be the total amount of sales to meet his goal?

(A) 9,050

(B) 11,900

(C) 190,000

(D) 250,000

16. A carpet costs \$10 per square meter (m^2). To the nearest dollar, how much would it cost to buy a carpet for 300 square feet(ft^2)? (1 meter = 3.28 feet)

17. The speed of light in water is 2.997×10^8 meters per second. which of the following closest to the speed of light in water, in kilometers per hour?
(A) 83.25
(B) 83,250
(C) 1.79×10^7
(D) 1.079×10^9

18. In a community garden, 31.9 million grams of tomato and 11.5 million grams of pepper were harvested. Of those plants, 8.8% of the tomato and 27.7% of the pepper were lost to late blight. The amount of discarded pepper was approximately what percent greater than that of discarded tomato?
(A) 13.5%
(B) 14.7%
(C) 15.1%
(D) 18.9%

19. Maria drives at an average rate of 45 miles per hour to a dentist, which takes 40 minutes to get there. If she drives 60 miles per hour, how many fewer minutes will it take her to travel the same distance?

20. 10.5 miles per gallon is approximately equal to how many kilometers per liter? (1 kilometer=0.62 miles and 1 gallon =3.79 liters)

(A) 1.72
(B) 4.47
(C) 24.67
(D) 64.19

21. An airplane flies at an average speed of 450 miles per hour and uses approximately 8 gallons of fuel per mile flown. If the plane can hold 60,000 gallons of fuel at the beginning of a trip, how many gallons of fuel are expected to remain in the fuel 14 hours after the trip begins?

22. The current density in a wire is defined as the current, in milliamperes(mA), flowing through the wire divided by the cross-sectional area of the wire in square millimeters. What is the current density,in milliamperes per square millimeter, in a battery with a cross-sectional area of 8 square millimeters when a current of 45 milliamperes flows?

(A) 0.178

(B) 5.625

(C) 45

(D) 360

23. A cooking pot has a density of 8.96 grams per cubic inch. An online company sells a cooking pot at a price of $0.05 per gram. What is the selling price, in dollars per cubic inch, for a cooking pot purchased from this company?

Number	Answer	Number	Answer	Number	Answer	Number	Answer
1	D	7	B	13	C	19	10
2	B	8	B	14	13.5	20	B
3	15	9	4340	15	C	21	9600
4	189	10	B	16	279	22	B
5	320	11	1625	17	D	23	0.448=.448
6	C	12	D	18	A		

4.5 Explanation

1. **(D)**

   $A = 1.3B, B = 0.85C, B = \frac{x}{1.3}, C = \frac{B}{0.85} = \frac{x}{(0.85)(1.3)} = \frac{x}{1.105}$

   $A + B + C = x + \frac{x}{1.3} + \frac{x}{1.105}$

2. **(B)**

   The number of tourists in country A in 2020 is 1.2(120,0000).

   Let B be the number of tourists in country B in 2000.

   Then, the number of tourists in country B in 2020 is $0.9B$

   $1.2(120,000) = 0.9B$

   $B = 160,000$

   The difference is $160,000 - 120,000 = 40,000$

3. **15**

   $\frac{2325}{15,500} \times 100 = 15\%$

4. **189**

   $0.92 \times 205 = 188.6 \approx 189$

5. **320**

   Let T be the total number of elk.

   $480 = 0.6T$

   $T = 800$

   Then, the number of male elk is $800 - 480 = 320$.

6. **(C)**

   $x(1 - \frac{p}{100}) = n$

   $x(\frac{100-p}{100}) = n$

   $x = \frac{100n}{100-p}$

7. **(B)**

The growth factor is $1 + 0.3$ every year for 12 years.

$x(1.03)^{12}$

8. **(B)**

Let x be the origianl price of the meals.

$x(0.8)(1.15) = 140.75$

$x = 152.989$

9. **4340**

$28,000(0.05) + (70,000 - 28,000)(0.07) = 4340$

10. **(B)**

$\frac{80(0.45)+30(0.51)}{80+30} = \frac{51.3}{110} = 0.466 = 46\%$

11. **1625**

Let x be the cost of the square sod with a side length 75 ft.

$130 : 450 = x : 5625$

$450x = 130 \cdot 5625$

$x = 1625$

12. **(D)**

$I = kE$

$4I = k(4E)$

13. **(C)**

Number of patents applied for in North America : Number of patents applied worldwide is $1 : 5$.

$200 : W = 1 : 5$

$W = 1,000$ thousands. $N = 200$ thousands. The number of patents applied for outside of North America is 800 thousands.

14. **13.5**

$30 : \frac{9}{4} = 15 \times 12 : x$

$x = 13.5$

15. **(C)**

Let s be the total amount of sales this month.

$3000 + (0.05)s = 12,500$

$s = 190,000$

16. **279**

    $1\text{m}^2 = (3.28)^2\text{ft}^2$

    $\frac{300}{(3.28)^2} \times 10 = 278.85 \approx 279$

17. **(D)**

    $\frac{2.997 \times 10^8}{10^3} \times 3,600 = 1,078,920,000 \approx 1.079 \times 10^9$

18. **(A)**

    The amount of discarded tomato is $0.088(31.9) = 2.8072$

    The amount of discarded pepper is $0.277(11.5) = 3.1855$

    $\frac{3.1855 - 2.8072}{2.8072} \times 100 = 13.476$

19. **10**

    Distance in miles is $45 \times \frac{40}{60} = 30\text{miles}$

    Time it took in 60 miles per hour is 30 minutes.

    $40 - 30 = 10\text{minutes}$

20. **(B)**

    $10.5\text{m/g} = \frac{10.5}{0.62}\text{km/g} = \frac{10.5}{(0.62)(3.79)}\text{km/l} = 4.46846$

21. **9600**

    speed=450 miles per hour

    rate=8 gallons per mile

    The remaining fuel is $60,000 - 8 \times 450 \times 14 = 9600$

22. **(B)**

    Density $= \frac{\text{Current}}{\text{Area}} = \frac{45}{8} = 5.625$

23. **0.448=.448**

    $0.05 \times 8.96 = 0.448$ dollars per cubic inch

Chapter 5

Word Problem

5.1 3 Types of Word Problems

3 Types of Word Problems

1. **System of linear equations**

   How many cubic centimeters (cm^3) of a solution that is 20% acid must you add to another solution that is 45% acid to produce 100 cm^3 of a solution that is 30% acid?

2. **Quadratic function modeling**

   When a basketball team charges \$4 per ticket, the average attendance is 5,000. For each \$0.1 increase in the ticket price, average attendance decreases by 20 people. What should the ticket price be to maximize revenue?

3. **Exponential function modeling**

   Mammalian cells grow in liquid culture. At the beginning of each day, there were twice as many cells in the culture as the preceding day. If there are 2,340 cells at the beginning of day 1, how many cells, y will be in the culture at the beginning of day x?

   (A) $y = 2340 + 2x$

   (B) $y = 2340 + 2(x-1)$

   (C) $y = 2340(2)^x$

   (D) $y = 1170(2)^x$

Explanation

1. System of linear equations

Let $x =$ the volume of 20% solution and $y =$ the volume of 45% solution.

$$0.2x + 0.45y = 100(0.3) = 30$$

$$x + y = 100$$

$$x = 60$$

Answer: 60

2. Quadratic function modeling

Let $x =$ the number of $0.1 increases. Then, the price of a ticket is $4 + 0.1x$ and the number of attendance is $5000 - 20x$.

$R(x) = (4 + 0.1x)(5000 - 20x)$

The maximum revenue occurs at $x = 105$ The price of a ticket is $14.5.

Answer: 14.5

3. Exponential function modeling

The number of cells grows by the factor of 2 every day. It is an exponential function and it passes through $(1, 2, 340)$.

Answer: (D)

5.2 Chapter 5 from A to Z Problems

1. A rectangle is four times as long as it is wide. If it has an area of 36 square inches, what is the length of the rectangle?

2. Kana burns about 5.3 calories per minute riding her bike and about 6.4 calories per minute jogging. If Kana spends 6 hours per week biking and jogging and burns a total of 1941 calories. How many minutes does she spend jogging?

3. The cranberry apple punch that contains 10% real juice is made by mixing x gallons of a cranberry drink with y gallons of an apple drink. The cranberry drink contains 12.5% real juice, and the apple drink contains 7% real juice. What volume, in gallons, of a cranberry drink will be needed if 30 gallons of an apple drink is used?

4. The half-life of strontium-90 is 28 years. If a 50-mg sample of a strontium-90 decays, how much of the sample will remain after 80 years, to the nearest tenth?

5. A buret is a tool designed to transfer precise amounts of liquid. A partially filled beaker containing 20 milliliters of a solution is placed under a leaky buret that produces one 0.05-milliliter drop of the solution every 10 seconds. Until the beaker is full, which of the following can be used to represent the volume, v in milliliters, of the solution in the beaker s seconds after it is placed under the buret?

(A) $v(s) = 0.5s + 20$
(B) $v(s) = 0.05s + 20$
(C) $v(s) = 0.005s + 20$
(D) $v(s) = -0.05s + 20$

6. Patrick measured the temperature of a cup of coffee placed in a room with a constant temperature of 70 degrees Fahrenheit ($°F$). The coffee cup's temperature was $200°F$ when it started cooling. After 10 minutes, the temperature of the coffee was $150°F$. Which of the following functions best models the temperature $F(m)$, in degrees Fahrenheit, of Patrick's coffee m minutes after it started cooling?

(A) $F(m) = 70(1.079)^m$
(B) $F(m) = 200(0.85)^m$
(C) $F(m) = (200 - 70)(0.9526)^m$
(D) $F(m) = 70 + 130(0.9526)^m$

7. James bought buttons and posters to give away for his campaign. He spent a total of $108 on b buttons and p posters. Each button costs $0.30, and each poster costs $0.15. The ratio of buttons to posters was 7 to 1. Which system of equations represents this situation, where b is the number of buttons, and p is the number of posters?

(A) $\begin{cases} 0.30b + 0.15p = 108 \\ 7b = p \end{cases}$

(B) $\begin{cases} 0.30b + 0.15p = 108 \\ 7p = b \end{cases}$

(C) $\begin{cases} 0.30b + 0.15p = 108 \\ 7bp = 1 \end{cases}$

(D) $\begin{cases} 0.30b + 0.15p = 108 \\ bp = 7 \end{cases}$

8. The president of an economics club found that when the price of a T-shirt sold for $18, a total of 60 T-shirts sold each day. For every $1 the price of the shirt is reduced, 10 additional T-shirt will be sold. Which equation models the amount collected R, in dollars, from T-shirt sales each day, where x is the number of $1 price decreases?

(A) $R = (18 + x)(60 + 10x)$
(B) $R = (18 - x)(60 + 10x)$
(C) $R = (18 + x)(60 + x)$
(D) $R = (18 - x)(60 - x)$

9. Sarah attends a graduate school that charges $1,200 per credit plus $1,350 each month for room and board. If Sarah takes x credits in a 6-month term, which of the following represents the total amount y, in dollars, that Sarah is charged for credits and room and board for the 6-month period?

(A) $y = 1,200x + 8,100$

(B) $y = 7,200x + 1,350$

(C) $y = 1,200x + 1,350$

(D) $y = 7,200x + 8,100$

10. In 1960, the total yearly fish production in Peru was about 3.7 million tonnes, and the fish production in the United States was about 2.5 million tonnes. Between 1960 and 2020, fish production in Peru increased by about 485,000 tonnes per year, and fish production in the United States increased by about 550,000 tonnes per year. Let P represent the fish production in million tonnes, and t represent the number of years after 1960. Which of the following system of equations can be used to find the year when both productions had approximately the same?

(A) $\begin{cases} P = 3.7 + 0.485t \\ P = 2.5 + 0.55t \end{cases}$

(B) $\begin{cases} P = 3.7t + 0.485 \\ P = 2.5t + 0.55 \end{cases}$

(C) $\begin{cases} P = 3.7 + 485,000t \\ P = 2.5 + 550,000t \end{cases}$

(D) $\begin{cases} P = 3.7t + 485,000 \\ P = 2.5t + 550,000 \end{cases}$

11. In a tournament, each game is played between two teams and the winning team of each game advances to play a game in the next round, and the losing team is eliminated from the tournament. If 256 teams play the game in the first round, which of the following functions gives the number of teams, $T(n)$, remaining to play in the nth round of the tournament?

(A) $T(n) = 256(\frac{1}{2})^n$
(B) $T(n) = 256(\frac{1}{2})^{n-1}$
(C) $T(n) = 256(2)^n$
(D) $T(n) = 256(2)^{n-1}$

12. Seniors in Lexington High School go camping for a week. If 2 students share one tent, 4 additional tents will be needed to distribute tents to all of the students. If 4 students share one tent, 4 tents will not be used. How many seniors go camping?

13. Gene earns his regular pay of $14 per hour for up to 40 hours of work per week. For each hour over 40 hours of work per week, Gene is paid $1\frac{1}{2}$ times his regular pay. How much does Gene earn in a week in which he works 73 hours?

14. A rectangle has an area of 40 square meters. Its width is 3 meters greater than its length. If the width is x, which of the following equation represents the area of the rectangle in terms of x?

(A) $x^2 + 3x - 40 = 0$
(B) $x^2 - 3x - 40 = 0$
(C) $x^2 + 3x + 40 = 0$
(D) $x^2 + 3x + 40 = 0$

	Car Rental A	Car Rental B
Initial fee	$1.45	$1.20
Fee per mimute	$0.25	$0.00
Fee per mile	$0.75	$1.95

15. Mr. Moon will be traveling 30 miles from his house to the airport using one of two car rentals. If it takes 50 minutes to get to the airport, how much more Car Rental B charges than Car Rental A?

(A) $23.25

(B) $23.95

(C) $29.75

(D) $45.75

16. Two cars leave a gas station simultaneously, one traveling north and the other east. The northbound car travels 50 miles per hour, and the eastbound car travels 40 miles per hour. After 3 hours, how many miles apart are two cars traveling to the nearest miles?

17. A worker turned on a hose to fill a swimming pool with water at a constant rate. The pool was exactly half full 2 hours ago, and three-quarters of the water is in the pool now. At this rate, how many hours from now will it take to be completely filled with water?

18. A survey asked a class of 50 students how many pets they had. The number of students with one pet is 4 times the number of students, s, who have two or more pets. If 3 students have no pets, which of the following equations represents this situation?

(A) $\frac{1}{4}s + 3 = 50$
(B) $4s + 3 = 50$
(C) $5s + 3 = 50$
(D) $8s + 3 = 50$

19. A minor league hockey team has collected ticket sales data over the past year. When the price of the ticket was $65.00, zero seats were purchased. They predict that for each $1 decrease in the ticket price, 100 more tickets will be sold. When the revenue is $90,000, which of the following could be the number of seats sold?

(A) 20
(B) 45
(C) 3,250
(D) 4,500

5.2 ANSWERS							
Number	Answer	Number	Answer	Number	Answer	Number	Answer
1	12	7	B	13	1253	19	D
2	30	8	B	14	B		
3	36	9	A	15	A		
4	6.9	10	A	16	192		
5	C	11	B	17	2		
6	D	12	32	18	C		

5.3 Explanation

1. **12**

   $l = 4w, l = 36, 4w^2 = 36 \ w^2 = 9 \ w = 3, l = 12$

2. **30**

   $5.3b + 6.4j = 1941$

   $b + j = 360$

   $5.3(360 - j) + 6.4j = 1941$

   $-33 = -1.1j$

   $j = 30$

3. **36**

   $0.125x + 0.07(30) = 0.1(x + 30)$

   $0.025x = 0.9$

   $x = 36$

4. **6.9**

   The amount of strontium-90 is $50 \cdot \left(\frac{1}{2}\right)^{\frac{t}{28}}$

   $50\left(\frac{1}{2}\right)^{\frac{80}{28}} = 6.9$

5. **(C)**

   The buret produces one 0.005 milliliter drop of solution every second. The volume of the solution in the beaker is increasing by the rate of 0.005 milliliter every second.

   $v(s) = 20 + 0.005s$

6. **(D)**

   The function satisfies $(m, F) = (10, 150)$ and $(m, F) = (0, 200)$.

   (D) $F(m) = 70 + 130(0.9526)^m$ passes through $(0, 150)$ and $(10, 149.9924)$.

7. **(B)**

$0.3b + 0.15p = 108$

$b : p = 7 : 1$

$b = 7p$

8. **(B)**

Let x be the number of $1 decreases in the price of the shirt. Since the revenue is the product of price of T-shirt and the number of T-shirts, $R(x) = (18 - x)(60 + 10x)$.

9. **(A)**

$y = 1,200x + 1350 \cdot 6 = 1,200x + 8,100$

10. **(A)**

P represents the number of fishes in million tonnes.

$P = 3.7 + \frac{485,000}{1,000,000}t = 3.7 + 0.485t$

$P = 2.5 + \frac{550,000}{1,000,000}t = 2.5 + 0.55t$

11. **(B)**

$T(n)$ satisfies $(1, 256)$.

The decay factor is $\frac{1}{2}$.

$T(n) = 256 \cdot (\frac{1}{2})^{n-1}$

12. **32**

Let s be the number of students and t be the number of tents.

$s = 2(t + 4)$

$s = 4(t - 4)$

$2t + 8 = 4t - 16$

$t = 12$

$s = 32$

13. **1253**

$14(40) + 14 \cdot \frac{3}{2}(73 - 40) = 1253$

14. **(B)**

If the width is x, the length is $x - 3$.

Area of the rectangle is $x(x - 3) = x^2 - 3x$

$x^2 - 3x = 40$

$x^2 - 3x - 40 = 0$

15. **(A)**

Rental A: $1.45 + 0.25(50) + 0.75(30) = 36.45$

Rental B: $1.2 + 1.95(30) = 59.7$

$59.7 - 36.45 = 23.25$

16. **192**

The northbound car travels 150 miles and the eastbound car travels 120 miles. The distance between them is $\sqrt{120^2 + 150^2} = 192$.

17. **2**

Since the volume is increasing at a constant rate, it is a linear function.

$m = \frac{\frac{3}{4} - \frac{1}{2}}{0 - (-2)} = \frac{1 - \frac{3}{4}}{x - 0}$

$x = 2$

18. **(C)**

The number of students with one pet is $4s$.

The number of students with 2 or more pets is s.

The number of students with no pets is 3.

$4s + s + 3 = 50$

$5s + 3 = 50$

19. **(D)**

Ticket sales is $(65 - x)(100x) = 90,000$. The possible x-values are $x = 20$ or $x = 45$. When $x = 45$, the number of seats sold is 4,500.

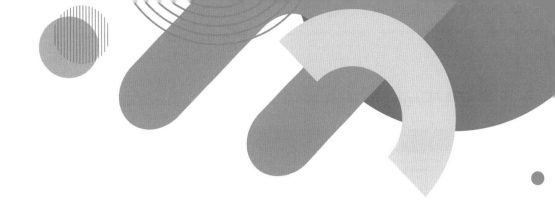

Chapter **6**

Probability

6.1 Probability

Probability

1. **Definition of Probability**

   Sample Space: The set of all possible outcomes.

   Event : Any outcome or set of outcomes in the sample space.

   $$P(E) = \frac{n(E)}{n(S)}$$

   $$= \frac{\text{The number of outcomes of event E}}{\text{The number of outcomes of sample space}}$$

   $$0 \le P(E) \le 1$$

2. **The Union of Two Events: A or B**

   If A and B are events in the same sample space, the probability of A or B is

   $$P(A \cup B) = P(A) + P(B) - P(A \cap B)$$

3. **The Intersection of Two Events: A and B**

   - If A and B are **independent events**, the probability of A and B is

   $$P(A \cap B) = P(A) \cdot P(B)$$

   - If A and B are **dependent events**, the probability of A and B is

   $$P(A \cap B) = P(A) \cdot P(B|A)$$

   $$= P(B) \cdot P(A|B)$$

4. **Complement Event**

   Let A be an event and A' be its complement.

   $$P(A') = 1 - P(A)$$

Conditional Probability

The probability that one event happens given that another event is already known to have happened is called a conditional probability.

1. The conditional probability of event B given that event A has happened.

$$P(B|A) = \frac{P(A \cap B)}{P(A)}$$

2. The conditional probability of event A given that event B has happened.

$$P(A|B) = \frac{P(A \cap B)}{P(B)}$$

6.1.1   Example

Explain the order of events and write the following statements in notation.

1. If an ATM is selected at random, what is the probability it belongs to a credit union?

2. What fraction of the plants that thrived received treatment A?

3. What is the percentage of adults aged 18-50 years who responded warm?

4. What is the probability that he has one or both ears pierced, given that the student that was chosen is male?

6.1.2 Example

At a sandwich shop, customers can order either a meat or a vegetarian sandwich on either white or wheat bread.

Type of bread	Meat	Vegetarian	Total
White	12	8	20
Wheat	16	14	30
Total	28	22	50

How many vegetarian ordered a sandwich on wheat bread?

(A) $\frac{4}{25}$

(B) $\frac{7}{25}$

(C) $\frac{4}{11}$

(D) $\frac{7}{11}$

6.1.3 Example

Marine biologists collected a sample of 400 adult sea turtles from Hammerhead Bay and estimated the age of each turtle in 2000 and 2010.

	2000	2010
Adult turtles under 30	220	195
Adult turtles 60 and older	59	78

If one of adult sea turtles was selected at random in the year 2000, what is the probability that the selected turtle is not under age 30? (Express your answer as a fraction or a decimal rounded to the nearest hundredth.)

6.1.4 Example

A group of 60 students took a field trip to an art museum. For their first guided tour, students were given a choice of 1 of 3 art exhibits. Of the 60 students, $\frac{2}{5}$ of the students chose Modern; the remaining students had an even distribution of the other two exhibits; Graffiti and Abstract. Each student chose exactly 1 exhibit. How many of the students chose Graffiti?

(A) 18
(B) 20
(C) 24
(D) 36

Explanation

6.1.1 Example

1. P(credit union| ATM)
2. P(treatment A| plants that thrived)
3. P(warm| adults aged 18-50 years)
4. P(one or both ears has pierced| male)

6.1.2 Example

$P(\text{wheat bread} \mid \text{vegetarian}) = \frac{14}{22} = \frac{7}{11}$

Answer: (D)

6.1.3 Example

$P(\text{turtles not under 30} \mid \text{adult sea turtles in 2000}) = \frac{(400-220)}{400} = \frac{180}{400} = \frac{9}{20} = 0.45$

Answer: $\frac{9}{20} = .45 = 0.45$

6.1.4 Example

$60 \times \frac{1}{2} \times (1 - \frac{2}{5}) = 18$

Answer: (A)

6.2 Chapter 6 from A to Z Problems

1. Eyespot is a disease in corn that is caused by a fungus. The table shows the relationship between the presence of eyespot in a cornfield and the assignment of fungicide.

	Fungicide assigned	Fungicide not assigned	Total
Eyespot observed	36	26	62
Eyespot not observed	10	14	24
Total	46	40	86

According to the table, rounded to the nearest tenth of a percent, what percentage of all cornfields where an eyespot was not observed needed fungicide?

Age(years)	Day 1	Day 2	Day 3	Total
16-25	16	24	40	80
26-45	54	48	53	155
46-60	65	23	12	100
Total	135	95	105	335

2. Data from a random sample of 335 sports car racing drivers in 2020 are listed above. The table indicates the number of drivers in 3 age groups and when each driver participated in the sports car racing over 3 days. No driver participated in the racing on two different days. If a driver is selected at random, what is the probability that the selected driver was in age 16-45, given that the driver raced on Day 2?

(A) $\frac{24}{95}$

(B) $\frac{72}{95}$

(C) $\frac{24}{335}$

(D) $\frac{72}{335}$

Department				
Faculty	Math	Physics	Biology	Total
Full-time	28	18	16	62
Part-time	x	y	36	93

3. The table shows the number of full-time and part-time faculty in a college's math, physics, and biology departments. The researchers found that a part-time math professor is as likely as a full-time math professor. What is the value of x?

Opinions on the Proposal				
	For	Against	Undecided	Total
Urban	526	980	95	1,601
Suburban	667	386	91	1,144
Rural	120	32	57	209
Total	1,313	1,398	243	2,954

4. The table above shows the results of a poll used to determine support for a state proposal. The results are categorized by demographics and opinion. If one person who responded to the poll is selected at random, which of the following statements results in the greatest value?

(A) The probability that the person is undecided, given that the person is from a rural area

(B) The probability that the person is undecided, given that the person is from an urban area

(C) The probability that the person is from a suburban area, given that the person is undecided

(D) The probability that the person is from a rural area, given that the person is undecided

5. Of the 200 different species of plant sold in a nursery, 120 species are annuals. Of the 120 species of annuals, 32 species grow best in full shade. If one of the species of plants sold in the nursery is selected at random, what is the probability of selecting an annual that does <u>not</u> grow best in full shade?

(A) $\frac{88}{200}$

(B) $\frac{88}{120}$

(C) $\frac{32}{200}$

(D) $\frac{32}{120}$

6. An integer from 1 to 150 is to be selected at random. What is the probability of selecting a number that is a multiple of 3 and less than or equal to 50? (Express your answer as a decimal or fraction, not as a percent.)

Alkalinity Class			
Depth class	Low	Medium	High
Shallow	87	61	209
Moderate	110	35	227
Deep	130	86	21

7. The table above shows the number of lakes in the United States classified by alkalinity and depth. If a lake has medium alkalinity, which of the following is closest to the probability that the lake also has a deep depth?

(A) 0.09

(B) 0.36

(C) 0.47

(D) 0.61

Department				
Country	Social science	Engineering	Language	Total
Mexico	65	35	15	115
Paris	75	64	26	165
Singapore	38	112	50	200
Guam	68	60	32	160
Total	246	271	123	640

8. The table above summarizes the survey results about travel destination preferences for a group of 640 university students of 3 different majors. Which of the following is closest to the difference between the percentage of students from Social science major who prefer Singapore or Guam and the percentage of students from Engineering who prefer Singapore or Guam?

   (A) 10%

   (B) 13%

   (C) 18%

   (D) 20%

9. In a study of penguins' breeding season, 240 King penguins and 160 Emperor penguins have been tagged. If 100 more Emperor penguins are tagged, how many more King penguins must be tagged so that $\frac{3}{5}$ of the total number of penguins in the study are King penguins?

	Types of Cases in millions		
States	Personal Injury	Immigration	Total
California		7,520	
Florida	970		

10. A lawyer is investigating the number of cases of personal injury and immigration in California and Florida in 2000. The study suggests that the number of immigration cases in each state was about 81% of total cases. The incomplete table above presents the information from this study, where the number of cases is given in millions. According to the table, how many more were the immigration cases, in millions, in California compared to Florida?

(A) 2,400
(B) 3,400
(C) 4,200
(D) 4,800

11. Each of the 100 distinct playing cards is 1 of 5 colors, and each has an integer value among $1 - 20$. There are 20 cards of each type of color; red, blue, green, black, and yellow. One of the 100 cards will be selected at random. What is the probability that the selected card will be numbered 7, given that the card is black?

(A) $\frac{1}{100}$
(B) $\frac{4}{100}$
(C) $\frac{1}{20}$
(D) $\frac{4}{20}$

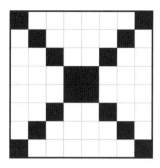

12. The figure shown is divided into 64 squares of equal area, where diagonal squares are shaded. If one of these squares is selected at random, what is the probability of selecting a square that is not shaded?

(A) 0.16
(B) 0.25
(C) 0.55
(D) 0.75

	Technology changes	No Technology change	Total
Faculty changes	30	80	110
No faculty chages	40	50	90
Total	70	130	200

13. In the table of the 2020 survey, 200 school administrators were asked about changes at their schools. What percentage of the administrators whose school made technology change given that the school faculty stays the same?

(A) 35%
(B) 44%
(C) 57%
(D) 100%

6.2 ANSWERS					
Number	Answer	Number	Answer	Number	Answer
1	41.7	7	C	13	B
2	B	8	D		
3	42	9	150		
4	C	10	B		
5	A	11	C		
6	$\frac{8}{75} = .1067 = 0.107$	12	D		

6.3 Explanation

1. **41.7**

   $P(\text{Fungicide assigned} \mid \text{Eyespot not observed}) = \frac{10}{24}$

   $\frac{10}{24} \times 100 = 41.7$

2. **(B)**

   $P(\text{16-45 years old} \mid \text{Day 2}) = \frac{24+48}{95} = \frac{72}{95}$

3. **42**

   $\frac{x}{93} = \frac{28}{62}$ $x = 42$

4. **(C)**

   (A) $\frac{57}{209} = 0.2727$

   (B) $\frac{95}{1601} = 0.059$

   (C) $\frac{91}{243} = 0.37448$

   (D) $\frac{57}{243} = 0.234567$

5. **(A)**

   P(annul that does not grow in full shade)$= \frac{120-32}{200} = \frac{88}{200}$

6. **$\frac{8}{75} = 0.107 = .1067$**

   The number of multiples of 3 less than 50 is 16.

   $\frac{16}{150} = \frac{8}{75}$

7. **(C)**

   $P(\text{medium alkalinity} \mid \text{Deep depth}) = \frac{86}{61+35+86} = \frac{86}{182} = 0.4725$

8. **(D)**

   $P(\text{Singapore or Guam} \mid \text{Social science}) = \frac{68+38}{246} = \frac{106}{246} = 0.4309$

   $P(\text{Singapore or Guam} \mid \text{Engineering}) = \frac{112+60}{271} = \frac{172}{271} = 0.6346$

   $100(0.6346 - 0.4309) = 20.37\%$

9. **150**

   Let x be the number of King penguins tagged later.

   $\frac{3}{5} = \frac{240+x}{160+100+240+x}$

   $1,500 + 3x = 1,200 + 5x$

   $x = 150$

10. **(B)**

The number of personal injury cases is 19% of the total number of cases in Florida.

$0.91 \times F = 970.$

$F = 5,105.263$

The number of immigration cases in Florida is 4135.263.

The difference is $7,520 - 4,135.263 = 3,384.737$.

11. **(C)**

$P(7 \mid \text{Black}) = \frac{1}{20}$

12. **(D)**

$\frac{64-16}{64} = \frac{48}{64} = \frac{6}{8} = 0.75$

13. **(B)**

$P(\text{Technology changes} \mid \text{No faculty changes}) = \frac{40}{90} \times 100 = 44.4\%$

Chapter 7

Geometry

7.1 Fundamental Geometric formulas

Definition 7: Geometric Formulas

1. Distance between 2 points, $P(x_1, y_1)$ and $Q(x_2, y_2)$ is

$$PQ = \sqrt{(x_2 - x_1)^2 + (y_2 - y_1)^2}$$

2. Midpoint between 2 points, $P(x_1, y_1)$ and $Q(x_2, y_2)$ is

$$M = \left(\frac{x_1 + x_2}{2}, \frac{y_1 + y_2}{2}\right)$$

3. Area of Triangle is

$$A = \frac{1}{2}bh$$

4. Area of Equilateral Triangle is

$$A = \frac{\sqrt{3}}{4}s^2$$

5. Area of Trapezoid is

$$A = \frac{h(b_1 + b_2)}{2}$$

6. Area and Circumference of Circle are

$$A = \pi r^2$$
$$C = 2\pi r$$

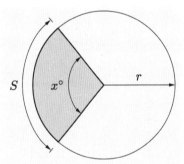

- Length of arc $S = 2\pi r \times \frac{x^\circ}{360^\circ}$
- Area of Sector $A = \pi r^2 \times \frac{x^\circ}{360^\circ}$

Figure 7.1

1. Surface area and Volume of Cube are

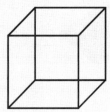

- $A = 6s^2$
- $V = s^3$

Figure 7.2

2. Surface area and Volume of Prism are

$$A = 2B + ph$$

where B=the area of base and p=the perimeter of base

$$V = Bh$$

3. Surface area and Volume of Cylinder are

$$A = 2\pi r^2 + 2\pi rh$$

$$V = \pi r^2 h$$

4. Surface area and Volume of Pyramids are

$$A = B + \frac{1}{2}pl$$

where B=the area of base, p=the perimeter of base and l=the slant height

$$V = \frac{1}{3}Bh$$

5. Surface area and Volume of Cone are

$$A = \pi r^2 + \pi rl$$
$$V = \frac{1}{3}\pi r^2 h$$

6. Surface area and Volume of Sphere are

$$A = 4\pi r^2$$
$$V = \frac{4}{3}\pi r^3$$

7.1.1 Example

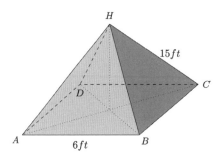

The regular pyramid shown above has a square base with side length 6ft. If a lateral edge of the pyramid has length 15ft, what is the volume of the pyramid, to the nearest tenth?

7.1.2 Example

A right cone A and Sphere B both have the same radius and volume. If the volume of the Sphere B is 50 cubic centimeters, what is the height of the cone A?

(A) 2.285

(B) 5.759

(C) 9.142

(D) 23.045

7.1.3   Example

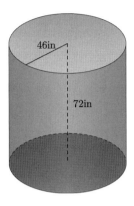

Jonathan is making cylindrical bales with a hay baler, as shown above. Each baler should be wrapped around it to hold its shape and a bale wrapper is plastic sheeting. How many <u>square feet</u> of plastic sheeting will he need to wrap five bales?

(A) $1,184.2$
(B) $14,210.5$
(C) $34,105.1$
(D) $170,525.6$

7.1.4   Example

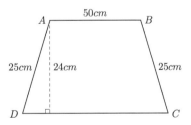

Quadrilateral $ABCD$ is an isosceles trapezoid. $AB = 50$cm, $AD = BC = 25$cm and the altitude is 24cm. What is the perimeter of the trapezoid?

7.1.5 Example

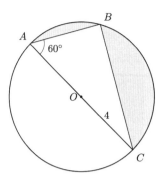

In the figure above, the circle has center O and the length of radius is 4. What is the area of the shaded region?

(A) $16(\sqrt{3} - \pi)$
(B) $16(\pi - \sqrt{3})$
(C) $8(\pi - \sqrt{3})$
(D) $8(2\pi - \sqrt{3})$

7.2 Angle Relationship

Definition 9: Angle Relationship

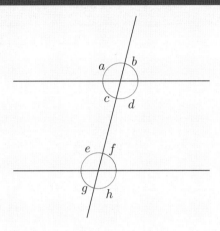

1. **Corresponding Angles Postulates**

   If two parallel lines are cut by a transversal, then the two pairs of corresponding angles are congruent.

   Ex) $\angle a \cong \angle e$, $\angle c \cong \angle g$ $\angle b \cong \angle f$, $\angle d \cong \angle h$

2. **Alternate Interior Angles Theorems**

   If two parallel lines are cut by a transversal, then the pairs of alternate interior angles are congruent.

   Ex) $\angle c \cong \angle f$, $\angle d \cong \angle e$

3. **Alternate Exterior Angles Theorems**

   If two parallel lines are cut by a transversal, then the pairs of alternate exterior angles are congruent.

   Ex) $\angle a \cong \angle h$, $\angle b \cong \angle g$

4. **Consecutive Interior Angles Theorems**

   If two parallel lines are cut by a transversal, then the pairs of consecutive interior angles are supplementary.

   Ex) $m\angle c + m\angle e = 180°$, $m\angle d + m\angle f = 180°$

7.3 Properties of Triangle

1. **Special Right Triangles**

$$1 : \sqrt{3} : 2$$

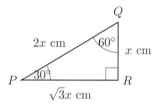

$$1 : 1 : \sqrt{2}$$

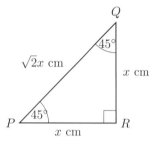

2. **Pythagorean Theorem**

    If a triangle is a right triangle, the square of the length of hypotenuse is equal to the sum of squares of the lengths and the legs.

    $$a^2 + b^2 = c^2$$

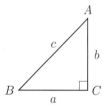

    Converse of the Pythagorean Theorem
    If $a^2 + b^2 = c^2$, then the triangle is a right triangle.
    Pythagorean Inequalities

    - If $c^2 > a^2 + b^2$, then the triangle is obtuse.
    - If $c^2 < a^2 + b^2$, then the triangle is acute.

3. **Triangle Inequality Theorem** The sum of the lengths of any two sides of a triangle is greater than the length of the third side.

$$a + b > c$$
$$b + c > a$$
$$c + a > b$$

Possible length of the third side, x is

$$|b - a| < x < b + a$$

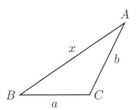

1. **Side-Side-Side(SSS) Postulate**

   If 3 pairs of corresponding sides of two triangles are congruent, then the two triangles are congruent.

   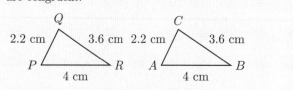

2. **Side-Angle-Side(SAS) Postulate**

   If 2 pairs of corresponding sides and included angles of two triangles are congruent, then the two triangles are congruent.

   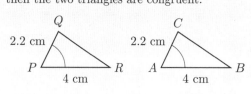

3. **Angle-Side-Angle(ASA) Postulate**

   If 2 pairs of corresponding angles and included sides of two triangles are congruent, then the two triangles are congruent.

   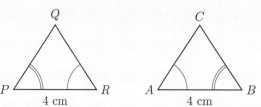

4. **Angle-Angle-Side(AAS) Theorem**

   If 2 pairs of corresponding angles and non-included sides of two triangles are congruent, then the two triangles are congruent.

   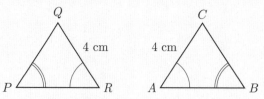

5. **Hypotenuse-Leg(HL) Theorem**

   If the hypotenuses and a pair of legs of two triangles are congruent, then the two triangles are congruent.

   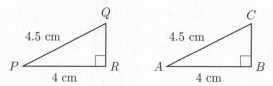

7.4 Similarity

1. **Angle-Angle (AA) Similarity Postulate**

   If two pairs of corresponding angles of two triangles are congruent, then the two triangles are similar.

   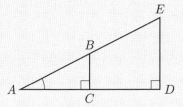

2. **Side-Side-Side (SSS) Similarity Theorem**

   If three pairs of corresponding sides of two triangles are proportional, then the two triangles are similar.

   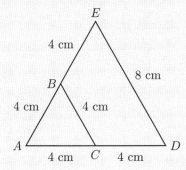

3. **Side-Angle-Side(SAS) Similarity Theorem**

   If two pairs of corresponding sides are proportional and included angles are congruent, then the two triangles are similar.

   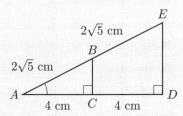

If two figures are similar, corresponding angles are congruent and side lengths are proportional.

Trigonometric ratios of corresponding angles are the same.

1. **Similar Figures**

   If the ratio of side lengths of two similar figures is $a : b$

   - ratio of perimeters is $a : b$
   - ratio of areas is $a^2 : b^2$
   - ratio of volumes is $a^3 : b^3$

2. **Proportionality Theorems**

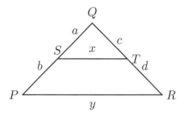

   -
   -
   -

   $$\frac{a}{b} = \frac{c}{d}$$

   $$\frac{a}{a+b} = \frac{x}{y}$$

   $$\frac{c}{c+d} = \frac{x}{y}$$

3. **Similar Right Triangles**

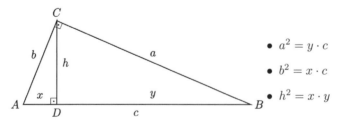

   - $a^2 = y \cdot c$
   - $b^2 = x \cdot c$
   - $h^2 = x \cdot y$

Figure 7.3

7.4.1   Example

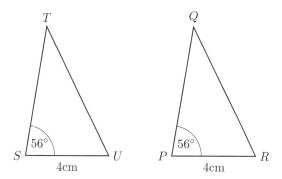

Triangle STU and triangle PQR each has an angle measuring $56°$ and a side length of 4cm, as shown above. For triangles STU and PQR, which additional piece of information is sufficient to prove that the triangle is congruent?

I. The length of side TU is equal to the length of side QR.

II. The length of side ST is equal to the length of side PQ.

(A) I is sufficient and II is not.
(B) II is sufficient but I is not.
(C) I is sufficient and II is sufficient.
(D) Neither I nor II is sufficient.

7.4.2   Example

The area of a rectangle is 65 square units. A second rectangle has side lengths 9 times those of the first rectangle. What is the area, in square units, of the second rectangle?

7.4.3 Example

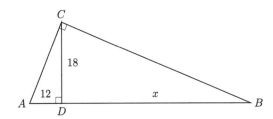

In the figure above, the point D lies on the hypotenuse of right triangle ABC. What is the length of line segment x?

7.5 Trigonometry

1. Basic trigonometric ratios

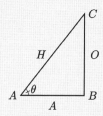

- $\sin \theta = \frac{O}{H}$
- $\cos \theta = \frac{A}{H}$
- $\tan \theta = \frac{O}{A}$

Figure 7.4

2. Trigonometric identities

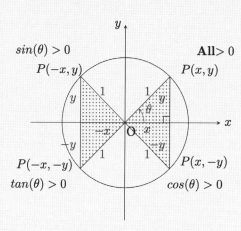

- $y = \sin(\theta)$
- $x = \cos(\theta)$
- $\tan(\theta) = \frac{\sin(\theta)}{\cos(\theta)}$
- $\sin(\theta) = \cos(\frac{\pi}{2} - \theta)$
- $\cos(\theta) = \sin(\frac{\pi}{2} - \theta)$
- $\sin(\pi - \theta) = \sin(\theta)$
- $\tan(\pi + \theta) = \tan(\theta)$
- $\cos(2\pi - \theta) = \cos(\theta)$
- $\sin(-\theta) = -\sin(\theta)$
- $\cos(-\theta) = \cos(\theta)$
- $\tan(-\theta) = -\tan(\theta)$

Figure 7.5

3. Degrees and Radians

$$\pi(rad) = 180°$$
$$1(rad) = \frac{180°}{\pi}$$
$$1° = \frac{\pi}{180}(rad)$$

θ	0	$\frac{\pi}{6}$	$\frac{\pi}{4}$	$\frac{\pi}{3}$	$\frac{\pi}{2}$	π	$\frac{3\pi}{2}$	2π
$\sin\theta$	0	$\frac{1}{2}$	$\frac{\sqrt{2}}{2}$	$\frac{\sqrt{3}}{2}$	1	0	-1	0
$\cos\theta$	1	$\frac{\sqrt{3}}{2}$	$\frac{\sqrt{2}}{2}$	$\frac{1}{2}$	0	-1	0	1
$\tan\theta$	0	$\frac{\sqrt{3}}{3}$	1	$\sqrt{3}$	UND	0	UND	0

Trigonometric Ratios

7.5.1 Example

Point A and point B lie on a circle with radius 1 meter, and the measure of arc $m\overset{\frown}{AB}$ is $\frac{7\pi}{12}$. What is the measure, in degrees, of this arc?

7.5.2 Example

Which of the following is the simplified form of $\frac{\cos(\frac{\pi}{2}-x)}{\sin(\frac{\pi}{2}-x)}$?

(A) $\frac{\cos(x)}{\sin(x)}$
(B) $\tan(\frac{\pi}{2}-x)$
(C) $\tan(x)$
(D) $-\tan(x)$

7.5.3 Example

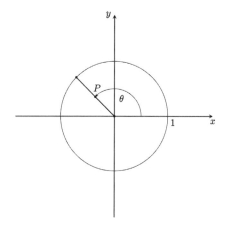

In the given unit circle, θ is an angle. If $\sin\theta = \frac{\sqrt{2}}{2}$, what is $\tan\theta$?

(A) -1
(B) $-\frac{\sqrt{2}}{2}$
(C) $\frac{\sqrt{2}}{2}$
(D) 1

7.5.4 Example

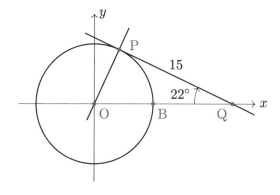

The tangent line $\overline{PQ}$ intersects the circle at P. The length of $\overline{PQ}$ is 15 centimeters, and the measure of $\angle PQO$ is $22°$. Which of the following value is the length, in centimeters, of $\overline{BQ}$?

(A) 7.85
(B) 10.12
(C) 16.18
(D) 33.98

7.6 Circle

Equation of a circle

A circle is the set of points that are equidistant from the center.

$$(x - h)^2 + (y - k)^2 = r^2$$

where $C(h, k)$ is the center of circle and r is the radius.

7.7 Properties of Circle

1.

Figure 7.6

- If the radius (or diameter) is perpendicular to the chord, it bisects the chord.

2.

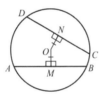

Figure 7.7

- Equal chords are equidistant from the center and chords equidistant from the center are equal in length.

3.

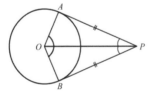

Figure 7.8

- A tangent line drawn to a circle is perpendicular to the radius (or diameter). Two tangents drawn to a circle from the same point outside the circle are equal in length.

4.

Figure 7.9

- The central angle of a circle is twice the inscribed angle.

5.

Figure 7.10

- If one side of a triangle in a circle is a diameter of the circle, then the triangle is a right triangle and the angle opposite the diameter is the right angle.

Explanation

7.1.1 Example

$$h^2 + (3\sqrt{2})^2 = 15^2$$

$$h = \sqrt{207}$$

$$V = \frac{1}{3} \cdot 36 \cdot \sqrt{207} = 172.6499 = 172.6$$

Answer: 172.6

7.1.2 Example

$$\frac{1}{3}\pi r^2 h = \frac{4}{3}\pi r^3 = 50$$

$$h = 4r$$

$$r^3 = \frac{3 \cdot 50}{4\pi}$$

$$r = 2.28539, h = 9.142$$

Answer: (C)

7.1.3 Example

1 feet $=$ 12 inches

1 square feet $=$ 144 square inches

Surface area $= \frac{5(2\pi \cdot 46^2 + 2\pi \cdot 46 \cdot 72)}{144} = 1184.205897$

Answer: (A)

7.1.4 Example

The length of upper base is $b_1 = 50$cm

The length of lower base is $b_2 = 7 + 50 + 7 = 64$cm

Perimeter of the trapezoid is $50 + 64 + 25 + 25 = 164$cm

Answer: 164

7.1.5 Example

Area$=\frac{1}{2}\pi 4^2 - \frac{4 \cdot 4\sqrt{3}}{2} = 8\pi - 8\sqrt{3}$

Answer: (C)

7.4.1 Example

I. SSA does not guarantee congruent triangles. This condition is not sufficient.

II. SAS guarantees congruent triangles. This condition is sufficient.

Answer: (B)

7.4.2 Example

The ratio of side lengths is $1 : 9$

Then, the ratio of areas is $1 : 81$

$1 : 81 = 65 : x$

$x = 5265$

Answer: 5265

7.4.3 Example

$18^2 = 12 \cdot x$

$x = 27$

Answer: 27

7.5.1 Example

π radian $= 180°$

$\frac{7\pi}{12} = \frac{7 \cdot 180}{12} = 105°$

Answer: 105

7.5.2 Example

$$\frac{\cos\left(\frac{\pi}{2} - x\right)}{\sin\left(\frac{\pi}{2} - x\right)} = \frac{\sin x}{\cos x} = \tan x$$

Answer: (C)

7.5.3 Example

$$\sin\theta = \frac{\sqrt{2}}{2}, \theta = \frac{3\pi}{4}$$

$$\tan\frac{3\pi}{4} = -1$$

Answer: (A)

7.5.4 Example

Let the radius be r.

$$\tan 22° = \frac{r}{15}$$

$$r = 15 \cdot \tan 22° = 6.06$$

$$OQ = \frac{15}{\cos 22°} = 16.178$$

$$BQ = 10.11$$

Answer: (B)

7.8 Chapter 7 from A to Z Problems–Part 1, Triangle and Polygon

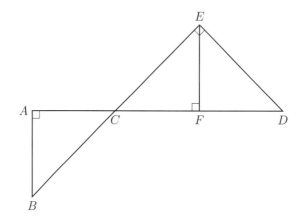

1. In the figure below, A, C, F, D are collinear and $B, C,$ and E are collinear. Line segment AD is perpendicular to line segments EF and AB. Which of the following statements could NOT be true from the given information?
   (A) $\overline{AB}$ is parallel to $\overline{EF}$.
   (B) $\angle ACB$ is congruent to $\angle FCE$.
   (C) $\angle ABC$ is congruent to $\angle FEC$.
   (D) $\overline{BC}$ is congruent to $\overline{CE}$.

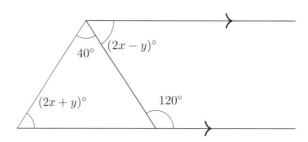

2. In the figure above, the given lines are parallel. What is the value of y?

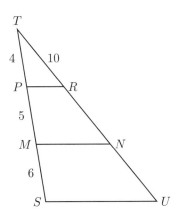

3. In the figure shown above, $\overline{PR}$ and $\overline{MN}$ are parallel to $\overline{SU}$. What is the length of segment $\overline{TU}$?

4. In the triangle STU, the measure of $\angle T$ is $90°$, $SU = 52$, and $UT = 48$. Triangle ABC is similar to triangle STU and each side of triangle ABC is $\frac{5}{2}$ the length of its corresponding side of triangle STU. What is the value of $\cos A$?

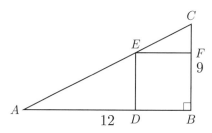

5. In the figure above, ABC is a right triangle and the quadrilateral $BDEF$ is a square. If $AB = 12$ and $BC = 9$, what is the area of the quadrilateral $BDEF$?

(A) 20.25

(B) 26.45

(C) 30.38

(D) 40.50

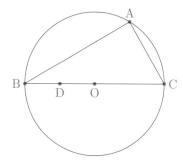

6. In the figure above, the circle has a center at O. If $\cos(\angle DAB) = 0.92$, what is the value of $\cos(\angle DAB) + \sin(\angle DAC)$?

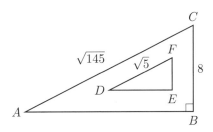

7. In the similar right triangles above, what is the area of the triangle, ΔDEF?
   (A) 1.241
   (B) 6.685
   (C) 193.866
   (D) 1044

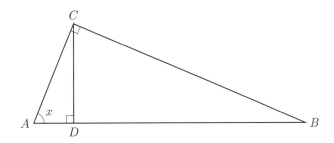

8. Triangle ABC is shown above. Which of the following is equal to $\cos(x)$?

   (A) $\frac{CD}{AC}$
   (B) $\frac{DB}{CB}$
   (C) $\frac{CD}{CB}$
   (D) $\frac{CD}{AD}$

9. Point P lies on a unit circle at coordinate $(1,0)$ and point O is at the center at coordinate $(0,0)$. Point Q also lies on the unit circle where $m\angle POQ = \frac{11\pi}{12}$. What is the coordinate of the point Q?

(A) $(-0.966, 0.259)$

(B) $(0.259, -0.966)$

(C) $(0.999, 0.050)$

(D)$(0.050, 0.999)$

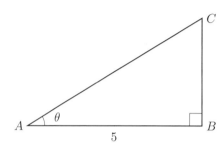

10. In right triangle ABC above, $AB = 5$. If the cosine of θ is $\frac{\sqrt{3}}{2}$, what is the length of BC?

(A) $\frac{5}{\sqrt{3}}$

(B) $\frac{5}{\sqrt{2}}$

(C) $5\sqrt{2}$

(D) $5\sqrt{3}$

11. In an equilateral triangle-based prism, base length is 10 inches and height is 8 inches. What is the surface area, in square inches, of the prism, to the nearest integer?

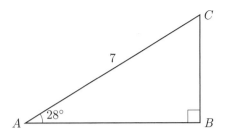

12. What is the area of triangle ABC above?

    (A) 10.156
    (B) 20.311
    (C) 59.104
    (D) 118.209

13. Given that $\sin\theta = \frac{5}{13}$ and $0 \le \theta \le 2\pi$, what are all possible values of $\cos\theta$?
    (A) $-\frac{5}{13}$
    (B) $-\frac{5}{13}$ and $\frac{5}{13}$
    (C) $\frac{12}{13}$
    (D) $-\frac{12}{13}$ and $\frac{12}{13}$

14. A regular hexagon is inscribed in a circle and the area of the regular hexagon is $108\sqrt{3}$ square inches. What is the radius of the circle, in inches?
    (A) $3\sqrt{2}$
    (B) $3\sqrt{6}$
    (C) $6\sqrt{2}$
    (D) 72

	Volume(m³)
Sphere A	324
Sphere B	12

15. The table gives the volumes, in cubic meters, of the two spheres. What is the ratio of the radius of Sphere B to the radius of Sphere A?

(A) $\frac{1}{27}$

(B) $\frac{1}{3}$

(C) 3

(D) 27

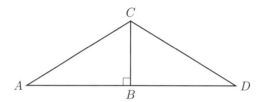

16. In the figure shown, segment CB is perpendicular to segment AD. Which additional piece of information is sufficient to determine whether triangle ABC is congruent to triangle DBC?

I. $AC = DC$

II. $\angle CAB \cong \angle CDB$

(A) I is sufficient, but II is not.

(B) II is sufficient, but I is not.

(C) I is sufficient, and II is sufficient.

(D) Neither I nor II is sufficient.

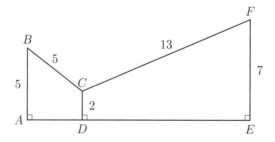

17. In the figure given above, line segments AB, CD, and EF are parallel. What is the total area of quadrilateral $ABCD$ and quadrilateral $CDEF$?

18. A pilot in a plane at an altitude of 16,480 feet observes that the angle of depression to a nearby airport is 6°. How many <u>miles</u> is the airport from the point on the ground directly below the plane? (1 mile = 5280 feet)

(A) 29.7

(B) 29.9

(C) 156,800

(D) 157,700

19. A company plans for renewing the size of a tuna can. The cylindrical can has a radius of 5 centimeters and a height of 30 centimeters. If the plan is increasing the radius by 20% and decreasing the height by 10%. What will be the increase, to nearest to 1%, in the volume of the tuna can?

(A) 8%

(B) 30%

(C) 32%

(D) 58%

7.9 Chapter 7 from A to Z Problems–Part 2. Circle

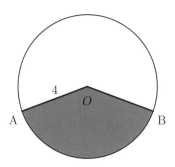

20. The circle shown above has center O and radius $\overline{AO}$. The radius $\overline{AO}$ of the circle has length 4 meters. If the area of shaded sector is 6π square meters. What is the measure of $\angle AOB$?

    (A) 105°

    (B) 120°

    (C) 135°

    (D) 160°

21. An arc of a circle measures $\frac{3}{2}$ radians. To nearest degree, what is the measure, in degrees, of this arc?

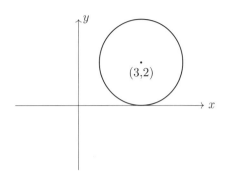

22. The circle has a center of $(3, 2)$ and is tangent to the x-axis. Which of the following is an equation of the circle shown above ?

    (A) $(x-3)^2 + (y-2)^2 = 9$

    (B) $(x-3)^2 + (y-2)^2 = 4$

    (C) $(x+3)^2 + (y+2)^2 = 9$

    (D) $(x+3)^2 + (y+2)^2 = 4$

23. In the xy-plane, the points $(-7, 10)$ and $(-5, 2)$ are the endpoints of a diameter of a circle. Which of the following is an equation of the circle?

    (A) $(x+6)^2 + (y+6)^2 = 68$

    (B) $(x+6)^2 + (y+6)^2 = 17$

    (C) $(x+6)^2 + (y-6)^2 = 68$

    (D) $(x+6)^2 + (y-6)^2 = 17$

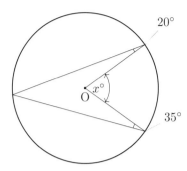

24. Point O is the center of the circle in the figure above. What is the value of x ?

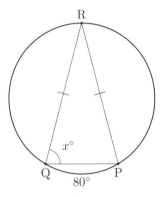

25. Triangle PQR is inscribed in the circle above. If the measure of arc $\overset{\frown}{QP}$ is $80°$, what is the measure of angle $x°$?

$$x^2 + y^2 - 10x + 5y = 0$$

26. In the xy-plane, the graph of the given equation is a circle. What is the radius of the circle?

(A) $\frac{\sqrt{5}}{2}$

(B) $\frac{5\sqrt{5}}{2}$

(C) $\frac{5}{2}$

(D) $\frac{125}{4}$

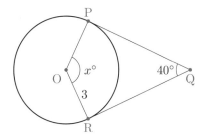

27. In the figure above, O is the center of the circle and the length of radius is 3. If the measure of angle $\angle PQR$, that is made of two tangent lines from the circle is 40°, what is the length of arc PR?

(A) $\frac{7\pi}{9}$

(B) $\frac{7\pi}{3}$

(C) $\frac{3\pi}{2}$

(D) π

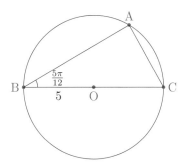

28. The figure above is a circle with radius of 5. If the measure of $\angle ABC$ is $\frac{5\pi}{12}$ radians, what is the area of the triangle ABC?

(A) 1.14

(B) 2.28

(C) 12.5

(D) 25

$$(x+3)^2 + (y+1)^2 = 26$$

29. The equation of a circle in the xy-plane is shown. The circle crosses the x-axis at points with coordinate $(a, 0)$. What is one possible value of a?

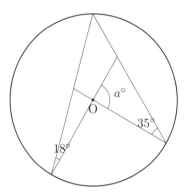

30. In the figure above, O is the center of the circle, What is the value of a ?

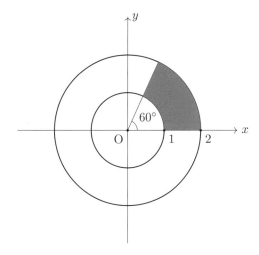

31. Concentric circles has center at O and thier radii are 1 and 2. What is the perimeter of the shaded region?

    (A) 1.04
    (B) 3.04
    (C) 3.14
    (D) 5.14

32. What is the diameter of the circle in the xy-plane that has center $(-1, 5)$ and passes through the origin, to the nearest tenth place?

Number	Answer	Number	Answer	Number	Answer	Number	Answer
1	D	7	A	13	D	19	B
2	10	8	C	14	C		
3	37.5	9	A	15	B		
4	$\frac{5}{13} = .3846 = 0.385$	10	A	16	C		
5	B	11	327	17	68		
6	1.84	12	A	18	A		

7.8 ANSWERS-Part 2. Circle					
Number	Answer	Number	Answer	Number	Answer
20	C	26	B	32	10.2
21	86	27	B		
22	B	28	C		
23	D	29	$2, -8$		
24	110	30	74		
25	70	31	D		

7.10 Explanation

1. **(D)**

   (A) Since EF and AD both are perpendicular to the same line, AD, $\overline{AB}$ is parallel to $\overline{EF}$.

   (B) Since $\angle ACB$ and $\angle FCE$ are vertical angles, they are congruent.

   (C) Since $\angle ABC$ and $\angle FEC$ are alternative interior angles, they are congruent.

   (D) It is not enough information.

2. **10**

   $2x - y = 60, 2x + y = 80$

   $x = 35, y = 10$

3. **37.5**

   $4 : 15 = 10 : TU$

   $TU = \frac{150}{4} = 37.5$

4. $\frac{5}{13} = \textbf{0.385} = \textbf{.3846}$

   $\cos A = \cos S = \frac{20}{52} = \frac{5}{13}$

5. **(B)**

   Let x be the length of the square $BDEF$.

   $x : 9 - x = 12 : 9$

   $x = \frac{36}{7}$

   Area$=(\frac{36}{7})^2 = 26.45$

6. **1.84**

   $\cos(x) = \sin(90° - x)$

   Let x be the measure of $\angle DAB$. Then, $90° - x$ is the measure of $\angle DAC$.

   $\cos(\angle DAB) = 0.92$ and $\sin(\angle DAC) = 0.92$

   $\cos(\angle DAB) + \sin(\angle DAC) = 1.84$

7. **(A)**

Ratio of lengths is $\sqrt{\frac{145}{5}} : 1$

Ratio of areas is $29 : 1$

$29 : 1 = 36 : A$

The area of $\triangle DEF$ is $\frac{36}{29} = 1.241$

8. **(C)**

$\cos x = \frac{AD}{AC} = \frac{CD}{CB}$

9. **(A)**

$(x, y) = (\cos \frac{11\pi}{12}, \sin \frac{11\pi}{12}) = (-0.9659, 0.2588)$

10. **(A)**

Since $\cos \theta = \frac{\sqrt{3}}{2}$, $\theta = 30°$.

$BC = \frac{5}{\sqrt{3}}$

11. **327**

Surface area $= 2 \cdot (\frac{\sqrt{3}}{4}) \cdot 10^2 + 30 \cdot 8 = 326.6 \approx 327$

12. **(A)**

$\frac{AB}{7} = \cos 28°$

$AB = 7 \cos 28°$

$BC = 7 \sin 28°$

Area $= \frac{1}{2} \times 6.18 \times 3.2863 = 10.1557$.

13. **(D)**

θ could lie in the first quadrant or the second quadrant.

Cosine value is positive in the first quadrant but negative in the second quadrant.

$\cos \theta = \pm \frac{12}{13}$

14. **(C)**

Let s be the side length of a regular hexagon.

s is the radius of the circle as well.

Area of hexagon is $6(\frac{\sqrt{3}}{4}s^2) = 108\sqrt{3}$

$s = \sqrt{72} = 6\sqrt{2}$

15. **(B)**

Ratio of volumes is $12 : 324 = 1 : 27$

Ratio of radii is $1 : 3$

16. **(C)**

I. By HL theorem, two triangles are congruent.

II. By AAS theorem, two triangles are congruent.

Both of them are sufficient to prove two triangles are congruent.

17. **68**

Area of trapezoid $ABCD$ is $\frac{4(2+5)}{2} = 14$

Area of trapezoid $CDEF$ is $\frac{12(2+7)}{2} = 54$

Total area is 68.

18. **(A)**

Let x be the distance between the airport and the point on the ground below the plane.

$\tan 6° = \frac{16,480}{x}$

$x = 156,796.7262\text{ft} = 29.69\text{mi}$

19. **(B)**

$(1.2)^2(0.9) = 1.296$

The percent increase is 29.6%

20. **(C)**

$6\pi = \pi \times 16 \times \frac{x°}{360°}$

$x = 135°$

21. **86**

$\frac{3}{2} \times \frac{180}{\pi} = 85.94 \approx 86$

22. **(B)**

Since the circle is tangent to the x-axis, the radius is 2. The center is $(h, k) = (3, 2)$

$(x - 3)^2 + (y - 2)^2 = 4$

23. **(D)**

The center of a circle is $(h, k) = (\frac{-7 + -5}{2}, \frac{10 + 2}{2}) = (-6, 6)$

$(x + 6)^2 + (y - 6)^2 = r^2$

$(-7 + 6)^2 + (10 - 6)^2 = 17$

$(x + 6)^2 + (y - 6)^2 = 17$

24. **110**

The inscribed angle is $55°$.

$x°$ is the central angle, which is the twice of the inscribed angle.

Thus, $x = 110$

25. **70**

Since the measure of arc QP is $80°$, $\angle QRP$ is $40°$.

$x° = \frac{180 - 40}{2} = 70$

26. **(B)**

$(x^2 - 10x + 25) + (y^2 + 5y + \frac{25}{4}) = 25 + \frac{25}{4} = \frac{125}{4}$

Radius is $\frac{5\sqrt{5}}{2}$.

27. **(B)**

Since a tangent line is perpendicular to the radius, $m\angle P = 90°$ and $m\angle R = 90°$. $x° = 140°$.

Length of arc is $2\pi \times 3 \times \frac{140}{360} = \frac{7\pi}{3}$.

28. **(C)**

$\frac{5\pi}{12} = 75°$

Area of triangle is $\frac{1}{2}AB \times AC = \frac{10 \cdot \cos 75° \, 10 \cdot \sin 75°}{2} = 12.5$

29. **$2, -8$**

$(x + 3)^2 + 1 = 26, (x + 3)^2 = 25 \; x + 3 = \pm 5 \; x = \pm 5 - 3 = -8, 2$

30. **74**

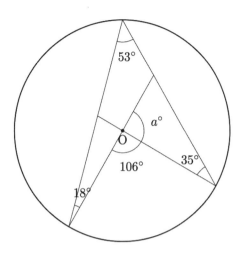

$a° = 180 = 106 = 74$

31. **(D)**

Perimeter$=2 + \frac{1}{6}(2\pi + 4\pi) = 2 + \pi = 5.14$

32. **10.2**

$(x + 1)^2 + (y - 5)^2 = 26$

Diameter is $2\sqrt{26} = 10.198 \approx 10.2$

Chapter **8**

Statistics

8.1 Describing Distributions with Numbers

Statistics

1. **Measure of central tendency**

   (a) **Mean**

   $$\bar{x} = \mu = \frac{\sum_{i=1}^{n} x_i}{n}$$

   $\bar{x}$ represents the sample mean.

   μ represents the population mean.

   (b) **Median**
   The middle value when the data values are arranged in order
   Location of median of n values is

   $$\frac{n+1}{2}$$

   (c) **Mode** The value that occurs with the greatest frequency.

2. **Spread**

   (a) **Range**
   The difference between the maximum data value and the minimum data value.

   $$\textbf{Range = Maximum - Minimum}$$

   (b) **Interquartile Range=IQR**
   The difference between the third quartile (Q_3) and the first quartile (Q_1) of the data.

   $$IQR = Q_3 - Q_1$$

   (c) **Standard Deviation**
   Standard deviation is the average distance of values from the mean.

   - Sample Standard Deviation,
   $$s_x = \sqrt{\sum \frac{(x-\bar{x})^2}{n-1}}$$
   - Population Standard Deviation,
   $$\sigma_x = \sqrt{\sum \frac{(x-\mu)^2}{n}}$$

Transforming the data

$$X = \{x_1, x_2, x_3, x_4, ..., x_n\}$$

$$Y = \{ax_1 + b, ax_2 + b, ax_3 + b, ax_4 + b, ..., ax_n + b\}$$

- New Mean

$$\bar{Y} = a\bar{X} + b$$

- New Median

$$Med_y = aMed_x + b$$

- New Range

$$Range_y = a(Range_x)$$

- New IQR

$$IQR_y = a(IQR_x)$$

- New Standard Deviation

$$\sigma_y = a(\sigma_x)$$

- **Outlier**

    An outlier is a value in a data set that significantly differs from other values. The inclusion of outliers in data sets can dramatically skew the summary statistics, which is why outliers are often removed from data sets.

Effect of the removal of an outlier on the mean

If a higher outlier is removed, the mean of the remaining values will decrease.

If a lower outlier is removed, the mean of the remaining values will increase.

8.1.1 Example

The prices, in thousands of dollars, of the 37 used cars at a certain car dealership are shown in the table below.

Price	Frequency
$9	1
$10	6
$11	7
$12	4
$13	2
$14	5
$15	7
$16	5

Which of the following statements best compares the mean and the median of the data shown in the frequency table?

(A) The median is 0.837 greater than the mean.
(B) The mean is 0.837 greater than the mean.
(C) The median is 0.162 greater than the mean.
(D) The median is equal to the mean.

8.1.2 Example

The total number of points scored by 7 teams in a curling tournament were collected. Later, it was found that reported scores of 7 teams were 5 points greater than the actual scores of 7 teams. Which of the following, statistics will remain unchanged if the total number of points are reported using the corrected scores?

(A) Mean
(B) Median
(C) Sum of the numbers
(D) Standard deviation

8.1.3 Example

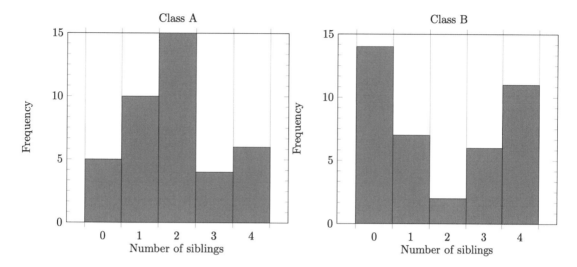

The histograms shown summarize the number of siblings of 40 students of each class, A and B. Which of the following statements best compares the ranges and standard deviations of the two classes?

(A) Class A has a greater range and a greater standard deviation than Class B.

(B) Class B has a greater range and a greater standard deviation than Class A.

(C) The range of Class A is equal to the range of Class B, and the standard deviation of Class A is greater than the standard deviation of Class B.

(D)The range of Class A is equal to the range of Class B, and the standard deviation of Class B is greater than the standard deviation of Class A.

8.1.4 Example

$$50, 66, 68, 70, 71, 71, 72, 76$$

Data set Q consists of 9 positive integers less than 80. The list shown above gives 8 of the integers from data set Q. The mean of these 8 integers is 68. If the mean of data set Q is an integer that is greater than 68, what is the value of the largest integer from data set Q?

8.2 Box and Whiskers Plot

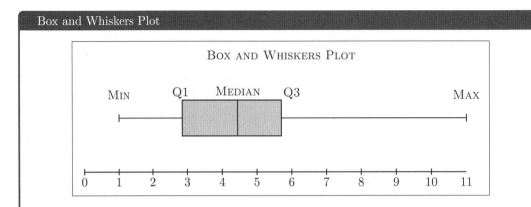

The five-number summary of a data set consists of the smallest observation, the first quartile, the median, the third quartile, and the largest observation, written in order from smallest to largest.

1. **First Quartile**, Q_1 = The median of lower half that are values less than the median.

2. **Third Quartile**, Q_3 = The median of upper half that are values greater than the median.

8.2.1   Example

Box and Whiskers Plot

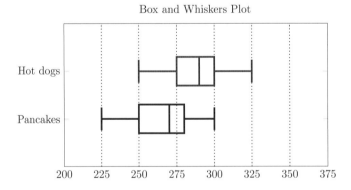

The box plot given above summarizes the sodium contents, in milligrams(mg) per serving, of 30 brands of pancakes and 30 brands of hot dogs. Based on the box plots, which of the following statements must be true?

I. The median of hot dogs brands is equal to the median of pancake brands.

II. The range of hot dogs brands is equal to the range of pancake brands.

(A) I only
(B) II only
(C) I and II
(D) Neither I nor II

8.3 Distribution, Population and Sample

Distributions

1. **Symmetric Distribution**

   If the right and left side of the graph are approximately mirror images of each other, the distribution is symmetric. In symmetric distribution,

   $$mean = median$$

2. **Right-Skewed Distribution**

   If the right side of the graph is much longer than the left side, the distribution is skewed to the right.

   $$mean > median$$

3. **Left-Skewed Distribution**

   If the left side of the graph is much longer then the right side, the distribution is skewed to the left.

   $$mean < median$$

1. **Population** Population is the entire group of individuals we want information about.

2. **Sample** Sample is a subset of individuals in the population from which we actually collect data.

3. **Sampling**

   (a) **Sampling Size**

   The larger the sample size, the more reliable result comes.

   (b) **Sampling Method: Random Selection**

   The sample should be randomly selected. When a sample is selected randomly, we can generalize the conclusion from the sample to the population.

   (c) **Representative Sample**

   The sample should represent the characteristics of population.

4. **C% confidence interval and Margin of error**

   A confidence interval gives an estimated range of values which is likely include an unknown population parameter. It's calculated from a sample taken from the population and is of the form:

   $$Statistic - ME \leq Population \ parameter \leq Statistic + ME$$

   $$\bar{x} - ME \leq \mu \leq \bar{x} + ME$$
   $$\hat{p} - ME \leq p \leq \hat{p} + ME$$

8.3.1 Example

A sample of 250 collegiate baseball players were selected from one region of the state. To investigate the mean annual salaries of collegiate baseball players, players completed a survey about their salaries. Which of the following is the largest population to which the results of the survey can be applied?

(A) The 250 selected collegiate baseball players
(B) All collegiate players in the region
(C) All collegiate baseball players in the region
(D) All collegiate baseball players in the state

8.3.2 Example

From a population of 1,200,000 people, 3,000 were chosen at random and surveyed whether they were optimistic about the economy. Based on the survey, it is estimated that 24% of people in the population were optimistic about the economy, with an associated margin of error of 3%. Based on these results, which of the following is a plausible value for the total number of people in the population who are optimistic about the economy?

(A) 600
(B) 750
(C) 255,000
(D) 330,000

8.3.3 Example

Sample	Mean	Margin of error
A	45	1.85
B	43	0.74

Two random samples, A and B, were selected from the same population to estimate the population mean, and the margin of error were calculated using the same method. Which of the following is the most appropriate reason that the margin of error for sample A is greater than the margin of error for sample B?

(A) The sample size of A is greater than the sample size of B.
(B) The sample size of A is less than the sample size of B.
(C) The sample size of A is equal to the sample size of B.
(D) The mean of sample A is greater than the mean of sample B.

8.4 Observational Study and Experiment

Observational Study	Experiment

Observational study observes individuals and measures variables of interest but does not attempt to influence the responses.
 	
1. **By observation**
    The variables must be observed, and the data must be collected through observation.
 	
2. **No interference**
    The researcher can only observe, and they cannot interfere with the study in any way.
	An Experiment deliberately imposes some treatment on individuals to measure their responses.
	
	1. **Comparison**
	    Use a design that compares two or more treatments.
	
	2. **Random assignment**
	    Use the impersonal chance to assign experimental units to treatments. Random assignment allows us to make inferences about **cause and effect** relationship between factors.
	
	3. **Control**
	    Keep other variables, such as lurking variables, that might affect the response.
	
	4. **Replication**
	    Use enough experimental units in each group so that any differences in the effects of the treatments can be distinguished from the chance difference between the groups.

Possible conclusion: Positive or Negative correlation (association) between two variables	**Possible conclusion:** Cause-effect relationship

8.4.1 Example

A study was conducted to determine whether oral insulin doses improve insulin absorption comparable to those produced by injections given through the skin for diabetic patients. 500 volunteers with diabetes were randomly assigned to receive treatments: pills and injections. Half of the patients received insulin pills, and the other half got insulin injections. The researcher concluded that absorbed insulin was significantly higher for those injected through the skin than for those who took a pill. Which of the following statements is correct?

(A) The insulin injection is likely to improve insulin absorption for these volunteers, and this conclusion can be generalized to all diabetic patients.

(B) The insulin injection is likely to improve insulin absorption for these volunteers, but it is not reasonable to generalize this conclusion to all diabetic patients.

(C) It is not reasonable to conclude that the insulin injection is better than the oral insulin dose.

(D) It is not possible to draw any conclusion from this experiment because volunteers were used.

8.5 Regression Analysis

Regression Analysis

A scatterplot is the most useful graph for displaying the relationship between two quantitative variables. The values of one variable appear on the horizontal axis, and the values of the other variable appear on the vertical axis. Each individual in the data appears as a point in the graph. We summarize the overall pattern by drawing a line of best fit or a curve of best fit.

- **Regression line or a line of best fit** A straight line that describes how y variable changes as x variable changes. The line is often to predict values of y for given values of x.

- **Extrapolation** The use of a regression line or curve to predict y variable for far beyond the domain of x variable. Such predictions are not accurate.

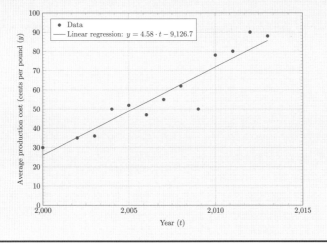

8.5.1 Example

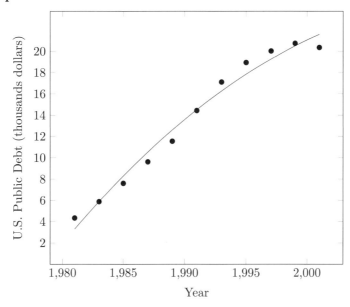

The scatterplot above shows the U.S. public debt per person, in thousands of dollars, in selected years for 11 people. The graph of a quadratic model for the data is also shown. For what fraction of the 11 people does the model overestimate the U.S. public debt?

Explanation

8.1.1 Example

Location of median is $\frac{37+1}{2} = 19$

Median is $13.

Mean is $\bar{x} = \frac{475}{37} = 12.3878$

The median is approximately 0.162 greater than the mean.

Answer: (C)

8.1.2 Example

When each data value decreases by the same number, standard deviation will not be changed.

Answer: (D)

8.1.3 Example

Range of Class A is 4 and Range of Class B is 4.

Since the data values in Class B spread out widely compared to distribution of Class A, standard deviation of Class B is greater than the standard deviation of Class A.

Answer: (D)

8.1.4 Example

$$\frac{68 \times 8 + x}{9} = 69$$

$x = 69 \cdot 9 - 544 = 77$

Answer: 77

8.2.1 Example

The median of hot dogs brands is approximately 285 and the median of pancakes brands is approximately 270. Thus, I statement is false.

The ranges of hot dogs brands and pancakes brands are both 75. Thus, II statement is true.

Answer: (B)

8.3.1 Example

The population is where the sample is selected from. The sample were selected from collegiate baseball players in one region of the state.

Answer: (C)

8.3.2 Example
The margin of error is applied to the sample statistic to create an interval in which the population statistic most likely falls. An estimate of 24% with a margin of error of 3% creates an interval of 24% ± 3%, or between 21% and 27%.
Thus, it's plausible that the percentage of people in the population who are optimistic about the economy is between 21% and 27%.
Let n be the total number of people in population who are optimistic about the economy.

$$1,200,000 \times 0.21 = 252,000 \leq n \leq 1,200,000 \times 0.27 = 324,000$$

Answer: (C)

8.3.3 Example
The greater the sample size, the smaller the margin of error would be.
Since the margin of error in sample A is greater than in sample B, the sample size of A is less than the sample size of B.

Answer: (B)

8.4.1 Example

The well-designed experiment was conducted on volunteers with random assignment, so the study results can be used to make conclusions about cause and effect on the population studied.

Since the amount of absorbed insulin was significantly higher for those who received insulin injections, we can conclude that insulin injection is likely to improve insulin absorption.

However, the sample was not selected randomly, so we cannot generalize the result to the population but only to the sample.

Answer: (B)

8.5.1 Example

The quadratic regression overestimates the U.S. public dept for actual data values below the curve. There are 4 points under the regression curve, so the fraction is $\frac{4}{11}$.

Answer: $\frac{4}{11} = 0.364 = .3636$

8.6 Chapter 8 from A to Z problems

1. A random sample of 50 mathematics teachers attending the conference was selected at a large conference of 3,000 teachers. Of those surveyed, 28% responded that they are willing to learn a new calculator. Based on the survey, which of the following is the best estimate of the total number of mathematics teachers intend to learn new calculators?

   (A) 14
   (B) 36
   (C) 840
   (D) 2,160

2. Of 102 customers at an electronics retail store, 63 stayed at the store for more than 0 minutes but less than 30 minutes. 30 people spent between 30 minutes and 1 hour, 7 people stayed between 1 hour and 1 hour and 30 minutes, and the remaining 2 customers spent 3 hours. Which of the following statements about the mean and median of shopping time is true?

   (A) The mean is greater than the median.
   (B) The median is greater than the mean.
   (C) The mean and the median are equal.
   (D) There is not enough information to determine whether the mean or the median is greater.

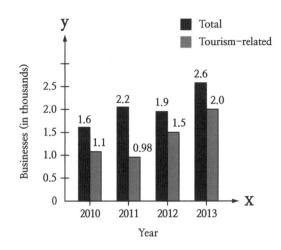

3. The bar graph shown summarizes the total number of businesses, in thousands, and the total number of tourism-related businesses, in thousands, in Hawaii for each of the four years. What is the mean number of businesses, in thousands, that are NOT related in tourism for four years?

(A) 0.68

(B) 1.40

(C) 6.35

(D) 3.47

4. A journal article reported the results of a study in which the mean time spent on homework per month was 55.7 hours, with an associated margin of error of 0.3 hours. A second study of the same population was conducted and included a larger random sample than the random sample from the first study. Based on the change in the sample size, which of the following is the most likely impact on the results?

(A) The margin of error from the second study is larger than the margin of error from the first study.

(B) The margin of error from the second study is smaller than the margin of error from the first study.

(C) The mean from the second study is larger than the mean from the first study.

(D) The mean from the second study is smaller than the mean from the first study.

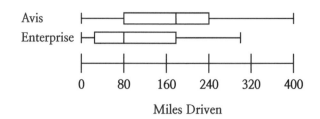

Miles Driven

5. There are car rental agencies in the city: Avis and Enterprise car rental agency. The box plots summarize the miles driven for one day of single-day car rentals at two agencies. Which of the following measures must be greater for miles driven by cars at Avis rental agency than for the miles driven by cars at Enterprise car rental agency?

I. The median

II. The range

(A) I only

(B) II only

(C) I and II

(D) Neither I nor II

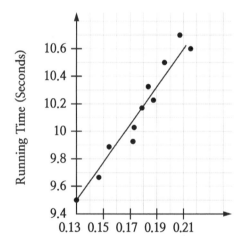

Reaction Time (Seconds)

6. Each data point on the scatterplot gives the running times for a 10-meter and the reaction times for 10 runners in a certain race. A line of best fit is also shown. Which of the following best approximates the equation for the line of best fit shown?

(A) $y = 7.71 + 13.8x$

(B) $y = 9.5 + 3.89x$

(C) $y = 7.71 - 13.8x$

(D) $y = 9.5 - 3.89x$

7. 10 rare auction items updated on an online auction website. If the value of each of the 10 items increases by $20 the next day, which of the following will be true?

(A) The new mean of the values will be $20 more than the previous mean, but the standard deviation will remain the same.

(B) The new mean of the values will remain the same, but the new standard deviation will be $20 more than the previous standard deviation.

(C) Both the new mean and standard deviation of the values will be $2 more than the previous mean and standard deviation.

(D) Neither the mean nor the standard deviation of the values will change.

Data set A	6	6	7	7	8	8	9	9	10	10
Data set B	16	16	17	17	18	18	19	19	20	20

8. The standard deviation of data set A is s and the standard deviation of data set B is t. Which of the following statements about the standard deviation of the data sets is true?

(A) $s > t$

(B) $s = t$

(C) $s < t$

(D) The relationship between s and r cannot be determined.

Plant name	Diameter (Km)
Mars	6,779
Jupiter	139,820
Earth	12,742
Saturn	116,460
Neptune	49,244
Venus	12,104
Mercury	4,879

9. The table shows the diameter, in kilometers km, for 7 Planets. What is the median length, in kilometer, of the diameters of 7 planets?

(A) 12,104

(B) 12,742

(C) 48,861

(D) 116,460

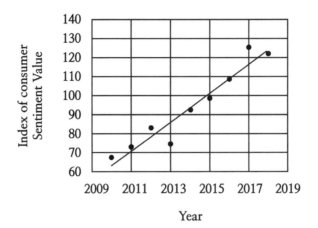

Year

10. The points on the scatterplot show the index of Consumer Sentiment values for US consumers each year from 2010 to 2018. A line of best fit for the data is also shown. For 2013, which of the following is a closest absolute difference between the actual Index of Consumer Sentiment value and the value predicted by the line of best fit?

(A) 5.6

(B) 12.0

(C) 20.2

(D) 75

11. Stephanie takes five tests, each worth a maximum of 100 points. Her scores on the first three tests are 78, 90, and 67. In order to average 80 for all five tests, what is the lowest score she could earn on one of the other two tests?

(A) 50

(B) 65

(C) 80

(D) 100

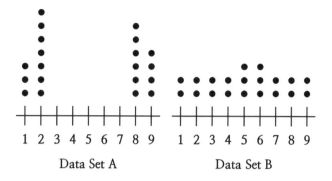

Data Set A   Data Set B

12. Which statement best compares the means of the two data sets shown?
(A) The mean of data set A is greater than the mean of data set B.
(B) The mean of data set A is equal to the mean of data set B.
(C) The mean of data set A is less than the mean of data set B.
(D) There is not enough information to compare the means.

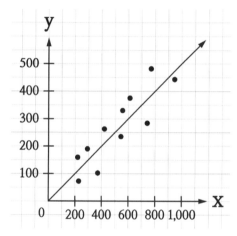

13. The scatterplot shows 11 data points, along with a line of best fit for the data. For how many of the data points does the line of best fit predict a y-value that is less than the actual y-value?

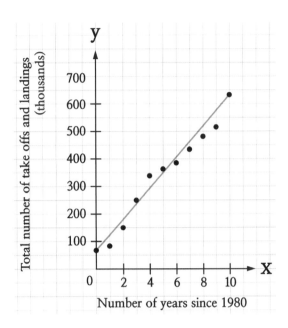

Total number of take offs and landings (thousands)

Number of years since 1980

14. The scatterplot shows the total number of takeoffs and landings in thousands at John F. Kennedy International Airport every year from 1980 through 1990. The equation of a line of best fit is $y = 56.77x + 59.375$. Which of the following is the best interpretation of the number 56.77 in this context?

(A) The increase in the total number of takeoffs and landings, in thousands, the model predicts at John F. Kennedy International Airport each year.

(B) The increase in the total number of takeoffs and landings, in thousands, the model predicts at John F. Kennedy International Airport from 1980 through 1990.

(C) The total number of takeoffs and landings, in thousands, the model predicts at John F. Kennedy International Airport for 1980.

(D) The total number of takeoffs and landings, in thousands, the model predicts at John F. Kennedy International Airport for 1990.

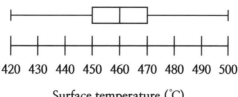

Surface temperature (˚C)

15. The box plot shows the surface temperature, in degrees Celsius($^\circ C$), each day for 40 solar days on Venus. Which of the following could be the number of solar days for which the surface temperature on Venus are at least $470^\circ C$?

(A) 10

(B) 20

(C) 25

(D) 30

16. A survey was conducted using a sample of history professors selected at random from the University of Michigan. The professors surveyed were asked to name the publishers of their current texts. What is the largest population on which the results at the survey can be generalized?

(A) All professors in the state of Michigan.

(B) All history professors in the state of Michigan.

(C) All history professors at the University of Michigan.

(D) All professors at the University of Michigan.

$$4.1, \ 4.1, \ 5.3, \ 5.3$$

17. The weights, in pounds, of four books are shown above. The standard deviation of these weights is 0.6 pound. Each book is packaged into a separate shipping box. Each empty box weights 4 pounds. What is the standard deviation of the weights of the four packages?

(A) 0.6 pound

(B) 2.4 pound

(C) 4.6 pound

(D) 16.6 pound

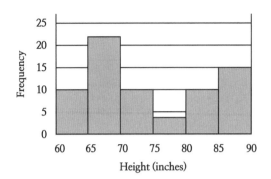

18. The histogram summarizes the distribution of height, inches, for a sample of 70 basketball players, where the first bar represents players who have a height of at least 60 inches but less than 65 inches, the second bar represents players who have a height of at least 65 inches but less than 70 inches, and so on. According to the histogram, how many basketball players in the sample have a height of at least 70 inches but less than 80 inches?

(A) 10

(B) 13

(C) 16

(D) 23

19. A student had a mean score of 68.8 for 10 tests. The student took an $11th$ test and had a score of 90, which is the highest score so far. If the teacher then dropped the lowest of the 11 scores, which of the following must be greater for the new set of 10 scores than for the original 10 scores?

I. The mean

II. The median

III. The standard deviation

(A) I only

(B) I and III

(C) II and III

(D) I, II and III

20. The members of a city council wanted to know the opinions of all city residents about conducting a spring music festival in the city square. The council surveyed a sample of 300 city residents who have purchased upcoming Symphony Orchestra. The survey showed that the majority of those sampled were in favor of the spring music festival. Which of the following is true about the city council's survey?

(A) It shows that the majority of city residents are in favor of the spring music festival.

(B) The survey sample is biased because it is not representative of all city residents.

(C) The survey sample should have included residents who live outside the city.

(D) The survey sample is biased because the survey sample should have more residents than 300.

21. Customers can park in the Ace Parking Garage for a maximum of 8 hours and the rate of the fee charged for parking depends on the length of time parked, as shown in the graph above. According to the graph, which of the following is NOT true?

(A) The fee for 2 hours of parking is twice the fee for 1 hour of parking.

(B) The fee for 4 hours of parking is four times the fee for 2 hours of parking.

(C) The rate of fee for 3 hours of parking is the same as the rate of fee for 3.5 hours of parking.

(D) The maximum fee for parking is 128 dollars.

22. A data set consists of 30 positive values, and the standard deviation of the data is b. A new data set is made by increasing each value in the original data set by 5 and then divided by 2. Which of the following represents the standard deviation of the new data set in terms of b ?

(A) $b + 5$

(B) $\frac{b}{2}$

(C) $\frac{b+5}{2}$

(D) $\frac{1}{2}(50b)$

Number	Answer	Number	Answer	Number	Answer	Number	Answer
1	C	7	A	13	6	19	A
2	A	8	B	14	A	20	B
3	A	9	B	15	A	21	A
4	B	10	B	16	C	22	B
5	C	11	B	17	A		
6	A	12	B	18	B		

8.7 Explanation

1. **(C)**
   $3,000 \times 0.28 = 840$

2. **(A)** Since the data has two extremely higher outliers, the distribution is skewed to the right. The mean would be greater than the median.

3. **(A)**
   $\bar{x} = \frac{(1.6-1.1)+(2.2-0.98)+(1.9-1.5)+(2.6-2)}{4} = 0.68$

4. **(B)**
   The second study has greater sample size than the first study, so the margin of error of the second study has smaller margin of error.

5. **(C)**
   The median of Avis is approximately 180 and median of Enterprise is 80.
   The range of Avis is 400 and the range of Enterprise is 310.
   Avis rental agency has greater median and range than those of Enterprise rental agency.

6. **(A)**
   $m = \frac{10.6-9.5}{0.21-0.13} = 13.75$
   The line of best fit passes through $(0.13, 9.5)$
   $y = 13.75x + 7.706 \approx 13.8x + 7.71$

7. **(A)**
   Adding $20 on each item will change the measure of the center, such as mean or median, but will not change shape or measures of spread, such as range, IQR or standard deviation. Therefore, the new mean will be increased by $20, but the standard deviation stays the same.

8. **(B)**
   Data set B values are increased by 10 from Data set A values. Since the same number is added to each data value in Data set B, there is no change in the measure of spread. Therefore, the standard deviations of A and B, $s = t$, are equal.

9. **(B)**
   The location of median is $\frac{7+1}{2} = 4$. The 4th length of the diameters is 12,742 in the ordered planets.

10. **(B)**

The actual Index of Consumer Sentiment Value in 2013 is approximately 75 and the predicted value by the line of best fit in 2013 is approximately 87.

The absolute difference of those is 12.0.

11. **(B)**

Assume that her unknown two test scores are x and y.

Then, $\frac{78+90+67+x+y}{5} = 80$.

$x + y = 165$.

In order to get a possible minimum score on one of the tests, we assume that the score on the other test is 100 points.

Then, the minimum score could be 65.

12. **(B)**

The mean of Data set A is $\frac{101}{20} = 5.05$. The mean of Data set B is $\frac{101}{20} = 5.05$.

Therefore, Both have equal means.

13. **6**

For the points above the line of best fit, y-values of those are underestimated by the line.

There are 6 points above the line, and those are underestimated.

14. **(A)**

56.77 stands for the slope of the linear regression. 56.77 represents the predicted increase in the total number of takeoffs and landings, in thousands, for each year at John F. Kennedy International Airport.

15. **(A)**

$470°C$ represents the third quartile. There are 25% data values above the third quartile.

$40 \times \frac{1}{4} = 10$

16. **(C)**

The largest population to which the survey results can be applied must be the group where the sample is selected.

Since the individuals in the sample are history professors at the University of Michigan, the largest population should be all history professors at the University of Michigan.

17. **(A)**

The weights of the four packages will be $8.1, 8.1, 9.3$, and 9.3. Since each data value is increased by 4, there is no change in the standard deviation.

The new standard deviation is 0.6.

18. **(B)**

10 basketball players have a height of at least 70 inches but less than 75 inches, and 3 basketball players have a height of at least 75 inches but less than 80 inches.

Therefore, 13 basketball players have a height of at least 70 inches but less than 80 inches.

19. **(A)**

The new mean must be greater than the original mean because the 11th test is the maximum value of the new data set, affecting the mean slightly higher than before.

We cannot determine the change in the median and the standard deviation because the exact values of 10 tests need to be given.

20. **(B)**

The sample is biased because the group surveyed does not represent all city residents. The sample is made of city residents who have purchased the upcoming Symphony Orchestra. They might be more interested in the music festival than others who did not buy the Symphony Orchestra ticket.

21. **(B)**

Even if adding a constant to a random variables does not affect the spread of data, multiplying random variables by a constant changes the spread of data, such as standard deviation. Therefore, the new standard deviation is $\frac{b}{2}$.

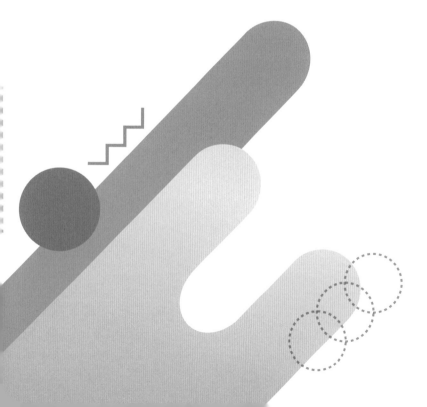

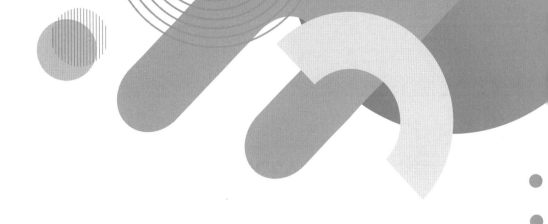

Chapter 9

Practice Tests

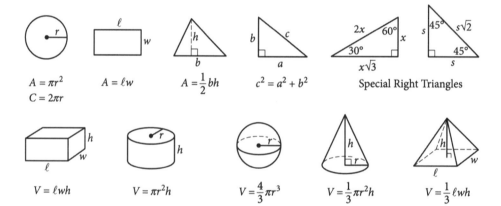

Module
1

PRACTICE TEST 1

Math

22 QUESTIONS

- The questions in this section address a number of important math skills.

- Use a calculator is permitted for all questions.

- Reference

The number of degrees of arc in a circle is 360.

The number of radians of arc in a circle is 2π.

The sum of the measures in degrees of the angles of a triangle is 180.

For multiple-choice questions, solve each problem,choose the correct answer from the choices provided, and then circle your answer in this book. Circle only answer for each question. If you change your mind, completely erase the circle. You will not get credit for questions with more than one answer circled, or for questions with no answers circled.

For student-produced response questions, solve each problem and write your answer next to or under the question in the test book as described below.

- Once you've written your answer, circle it clearly. You will not receive credit for anything written outside the circle, or for any questions with more than one circled answer.

- **If you find more than one correct answer**, write and circle only one answer.

- Your answer can be up to 5 characters for a **positive** answer and up to 6 characters (Including the negative sign) for a **negative** answer, but no more.

- If your answer is a **fraction** that is too long (over 5 characters for positive, 6 characters for negative), write the decimal equivalent.

- If your answer is a **decimal** that is too long (over 5 characters for positive, 6 characters for negative), truncate it or round at the fourth digit.

- If your answer is a **mixed number** (such as $3\frac{1}{2}$), write it as an improper fraction (7/2) or its decimal equivalent (3.5).

- Don't include **symbols** such as a percent sign, comma, or dollar sign in your circled answer.

Answer	Acceptable ways to enter answer	Unacceptable: will NOT receive credit
3.5	3.5 3.50 7/2	31/2 31/2
$\frac{2}{3}$	2/3 .6666 .6667 0.666 0.667	0.66 .66 0.67 .67
$-\frac{1}{3}$	$-\frac{1}{3}$ $-.3333$ -0.333	$-.33$ -0.33

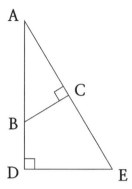

1. Triangle ABC and Triangle AED are right triangles. Which of the following is needed to determine whether triangle ABC is similar to triangle AED?

   (A) The measure of angle A

   (B) The measure of angle B and the measure of angle E

   (C) The length of side AE

   (D) No additional information is needed

2. Which of the following expression is equivalent to $\left(\sqrt{x} - \sqrt{2y}\right)^{\frac{2}{3}}$, where $x > 0$ and $y > 0$?

   (A) $(x - 2y)^3$

   (B) $\sqrt[3]{x - 2y}$

   (C) $\sqrt[3]{x - 2\sqrt{xy} + 2y}$

   (D) $\sqrt[3]{x - 2\sqrt{2xy} + 2y}$

$$3x + 2y = 90$$

3. The given equation represents the possible numbers of seeds collected from Black turtle beans and Pinto beans, where x is the number of Black turtle beans and y is the number of Pinto beans. Which of the following is the best interpretation of $3x$ in this context?

   (A) There are $3x$ seeds from two beans.

   (B) There are $3x$ seeds from Black turtle beans

   (C) There are $3x$ seeds from Pinto beans

   (D) Black turtle beans has $3x$ seeds per each bean.

$$18x + 6 = 3(3x + 2) + 8x$$

4. How many solutions does the given equation have?

   (A) Zero

   (B) Exactly one

   (C) Exactly two

   (D) Infinitely many

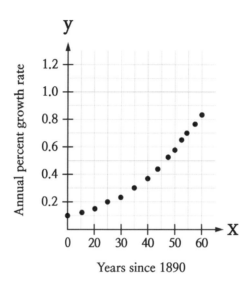

Years since 1890

5. The graph models the annual percent growth rate of the revenue of a particular marketing company from 1890 through 1950. Based on the model, which of the following is the best estimate of the annual percent growth rate of the revenue for 1920?

   (A) 0.11

   (B) 0.18

   (C) 0.24

   (D) 0.39

6. The function f is linear, and $f(3) = 11$. When the value of x increases by 4, the value of $f(x)$ increase by 3. Which of the following defines f?

(A) $y = \frac{4}{3}x + 5$

(B) $y = \frac{4}{3}x + 7$

(C) $y = \frac{3}{4}x + \frac{7}{4}$

(D) $y = \frac{3}{4}x + \frac{35}{4}$

$$y = 2,000,000\,(1.045)^x$$

7. The given equation models the age of the sediment in the valley of a canyon y, in years, where x is the number of 100 meters below the top of the canyon. The age of the sediment at the top of the canyon is $2,000,000$ years old. Which of the following equations best models the age of the sediment in the valley of a canyon m meters below the top of the canyon ?

(A) $y = \frac{2,000,000}{100}\,(1.045)^m$

(B) $y = 2,000,000\,(1.045)^{100m}$

(C) $y = 2,000,000\,(1.045)^{\frac{m}{100}}$

(D) $y = 2,000,000\left(\frac{1.045}{100}\right)^m$

$$(2x - 1)^2 - 10(2x - 1) + 25 = 0$$

8. What is the value of x ?

9. A quadratic function can be used to model the height y, in feet of a volleyball x seconds after it was hit by a player. According to the model, the volleyball was shot from a height of 5 feet and reached its maximum height of 6 feet 2.25 seconds after it was hit. How many seconds later would the volleyball be 5 feet again after it was shot?

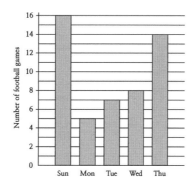

10. The bar graph shows the number of football games playing in a certain week. For these five days, what is the median number of games per day?

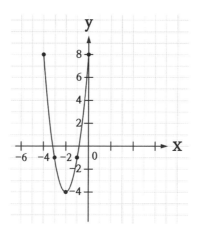

11. The graph of $y = 3(x+h)^2 + k$, where h and k are constants, is shown. What is the value of h?

(A) -4

(B) -2

(C) 2

(D) 8

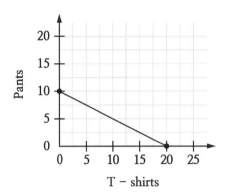

12. Hailey is buying T-shirts for $50 each and Pants for $100 each, including sales tax. The graph models the possible combinations of T-shirts and pants she can buy. What is the total amount of money she is spending on T-shirts and pants?

(A) $2,000

(B) $1,000

(C) $500

(D) $125

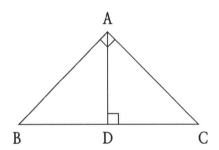

13. In the figure, $\triangle ABC$ is similar to $\triangle DBA$ and $\triangle DAC$. $AD = 10$ and $DC = 4$. What is the value of $\tan B$?

14. The two quantities y and x are related such that $y = 10$ when $x = 1$. When the value of x increases by 1, the value of y is multiplied by 4. Which of the following represents this relationship?

(A) $y = 10\,(4)^x$

(B) $y = 10\,(x - 1)^4$

(C) $y = 10\,(4)^{x-1}$

(D) $y = 10x^{\frac{1}{4}}$

15. A bakery sold c cupcakes during week days. The bakery sold 55% more cupcakes on weekend. How many cupcakes were sold on weekend, in terms of c?

(A) $0.55c$

(B) $0.155c$

(C) $1.55c$

(D) $155c$

16. A leopard leaps at an average speed 58 kilometers per hour. What is this speed in <u>meters per second</u>? (1 kilometer=1,000 meters)

(A) 16.1

(B) 161.1

(C) 208.8

(D) 966.7

$$y = -2(x - 1)(x + 3)(x - 4)$$

17. In the xy-plane, what is the value of y-coordinate of y-intercept?

    (A) -24

    (B) 0

    (C) 1

    (D) 4

18. In triangle ABC, which of the following statement guarantees that $\triangle ABC$ is an isosceles right triangle?

    I. $m\angle A = 45°$ and $m\angle B = 90°$
    II. $AB = BC$ and $AC = \sqrt{2}AB$

    (A) I is true, but II is false.
    (B) II is true, but I is false.
    (C) I is true, and II is true.
    (D) Neither I nor II is true.

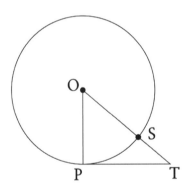

19. Point O is the center of the circle shown. Line PT is tangent to the circle at point P. The measure of minor arc $\overset{\frown}{PS}$ is $67°$ and the radius of the circle is 5. What is the length of segment OT?

    (A) 5.43

    (B) 8.66

    (C) 11.78

    (D) 12.80

20. Two fair number cubes with faces numbered $1, 2, 3, 4, 5$ and 6 are to be rolled together. What is the probability that product of number on the top faces is even?

(A) $\frac{1}{4}$

(B) $\frac{3}{4}$

(C) $\frac{10}{12}$

(D) $\frac{1}{2}$

State	Number of young adults who live with their parents (in thousands)
California	3,319
New Jersey	257
North California	223
Hawaii	146
New York	108
Colorado	103

21. The 2021 Population survey researched the number of young adults who live with their parents in six states. Suppose the number of young adults from California is excluded from the data set. What is the difference in the mean number of young adults between the original data set and the new data set to the nearest integer?

22. A sample of clay has a mass of 5.5 pounds and a volume of 0.011 cubic feet. What is the density, in pounds per cubic foot, of this clay?

PRACTICE TEST 1
Easy

Math
22 QUESTIONS

- The questions in this section address a number of important math skills.

- Use a calculator is permitted for all questions.

- Reference

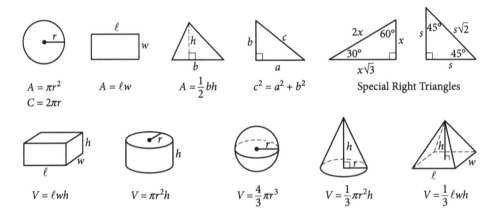

$A = \pi r^2$
$C = 2\pi r$

$A = \ell w$

$A = \frac{1}{2}bh$

$c^2 = a^2 + b^2$

Special Right Triangles

$V = \ell wh$

$V = \pi r^2 h$

$V = \frac{4}{3}\pi r^3$

$V = \frac{1}{3}\pi r^2 h$

$V = \frac{1}{3}\ell wh$

The number of degrees of arc in a circle is 360.

The number of radians of arc in a circle is 2π.

The sum of the measures in degrees of the angles of a triangle is 180.

For multiple-choice questions, solve each problem,choose the correct answer from the choices provided, and then circle your answer in this book. Circle only answer for each question. If you change your mind, completely erase the circle. You will not get credit for questions with more than one answer circled, or for questions with no answers circled.

For student-produced response questions, solve each problem and write your answer next to or under the question in the test book as described below.

- Once you've written your answer, circle it clearly. You will not receive credit for anything written outside the circle, or for any questions with more than one circled answer.

- **If you find more than one correct answer**, write and circle only one answer.

- Your answer can be up to 5 characters for a **positive** answer and up to 6 characters (Including the negative sign) for a **negative** answer, but no more.

- If your answer is a **fraction** that is too long (over 5 characters for positive, 6 characters for negative), write the decimal equivalent.

- If your answer is a **decimal** that is too long (over 5 characters for positive, 6 characters for negative), truncate it or round at the fourth digit.

- If your answer is a **mixed number** (such as $3\frac{1}{2}$), write it as an improper fraction (7/2) or its decimal equivalent (3.5).

- Don't include **symbols** such as a percent sign, comma, or dollar sign in your circled answer.

Answer	Acceptable ways to enter answer	Unacceptable: will NOT receive credit
3.5	3.5 3.50 7/2	31/2 31/2
$\frac{2}{3}$	2/3 .6666 .6667 0.666 0.667	0.66 .66 0.67 .67
$-\frac{1}{3}$	$-\frac{1}{3}$ $-.3333$ -0.333	$-.33$ -0.33

1.

$$4x + Cy = 3$$

$$3x + Cy = 5$$

In the system of equations above, C is a nonzero constant. If (x, y) is the solution to the system of equations, which of the following is (x, y), in terms of C ?

(A) $(-2, 11C)$

(B) $(11C, -2)$

(C) $\left(-2, \frac{11}{C}\right)$

(D) $\left(\frac{11}{C}, -2\right)$

2. In the xy-plane, an equation of Circle A is $(x-1)^2 + y^2 = 1$. Circle B is obtained by enlarging the circle by a factor of 2 about the center. Which of the following is an equation of circle B?

(A) $(2x-2)^2 + (2y)^2 = 4$

(B) $(x-1)^2 + y^2 = 4$

(C) $(x-1)^2 + y^2 = 2$

(D) $(x-2)^2 + y^2 = 2$

3.

$$(2x+9) - 10(2x+9)^2$$

Which of the following expressions is equivalent to the given expression?

(A) $(9-2x)(20x+89)$

(B) $(2x+9)(20x+89)$

(C) $-(2x+9)(20x+89)$

(D) $40x^2 + 358x + 801$

$$\frac{x^2\,(x-3) - 9(x-3)}{x^2 - 6x + 9}$$

4. If $x \neq 3$, which of the following expressions is equivalent to the given expression?

   (A) $x + 3$

   (B) $(x+3)\,(x-3)$

   (C) $(x-3)^2$

   (D) $\frac{1}{x-3}$

$$4x^2 - 1 = 0$$

5. What is the sum of the two solutions to the equation above?

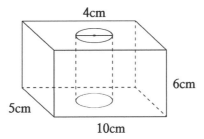

6. In the right rectangular prism, a cylindrical hole of a diameter of 4 centimeters is removed from its center, as shown in the figure. What is the volume, in cubic centimeters, of the prism after the cylinder was removed?

   (A) 225

   (B) 237

   (C) 262

   (D) 269

7. What is 170% of 2,000?

    (A) 140

    (B) 1,400

    (C) 3,400

    (D) 5,400

X	f(x)
0	100
5	40
10	16
15	6.4

8. For an exponential function, f, the table shows several values of x and their corresponding values of $f(x)$. Which of the following could define f?

    (A) $f(x) = 100(0.4)^{\frac{x}{5}}$

    (B) $f(x) = 100(0.6)^{\frac{x}{5}}$

    (C) $f(x) = 100(0.4)^{x}$

    (D) $f(x) = 100(0.6)^{x}$

9. A Ferris Wheel rotates 120 rotations per minutes. How many rotations does it complete in second?

10. The total yearly peanut production in the United States in 2016 is 3,700 thousand tonnes and increases by 485 thousand tonnes every year. In what year does the totally yearly peanut production in the United States to be 5,640 thousand tonnes?

    (A) 2019

    (B) 2020

    (C) 2024

    (D) 2067

11. The ratio of the lengths of a rectangle is 5 to 2. If the area of the rectangle is 90 square centimeters, what is the length, in centimeters, of the shorter side?

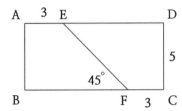

12. As shown above, $\overline{EF}$ divides rectangle $ABCD$ into 2 congruent trapezoids. The measure of $\angle BFE$ is $45°$. What is the area of rectangle $ABCD$?

(A) 45

(B) 48

(C) 55

(D) 110

13. In the xy-plane, which of the following functions is perpendicular to the linear function $3x = 4y + 2$?

(A) $y = -\frac{4}{3}x - 2$

(B) $y = -\frac{3}{4}x + 2$

(C) $y = \frac{3}{4}x - \frac{1}{2}$

(D) $y = \frac{4}{3}x + 1$

Velocity	Total distnace in meters
1	7
2	28
4	112

14. The table above shows the velocity, v in meters per second, of an object moving in a straight line and its corresponding total distance, d in meters. The function d, in miles is a quadratic function with respect to v. What is the total distance traveled of the object when its velocity is 3 meters per second?

(A) 14

(B) 56

(C) 63

(D) 70

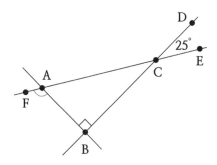

15. In the figure above, AB is perpendicular to BC and the measure of $\angle DCE$ is 25°. What is the measure of $\angle FAB$?

$$8.01,\ 6.92,\ 8.02,\ 7.15,\ 6.62,\ 6.14,\ 7.03$$

16. The given seven numbers represent the pH value of seawater from 7 samples. If the pH scale is greater than 7, seawater is alkaline. If one of the samples is selected at random, what is the probability of selecting a sample that is alkaline?

$$x = 5\left(\frac{y}{23}\right)^{\frac{3}{2}}$$

17. The given equation relates the positive numbers x and y. Which equation correctly express y in terms of x ?

(A) $y = \frac{23}{5}x^{\frac{2}{3}}$

(B) $y = \frac{23}{5}x^{\frac{3}{2}}$

(C) $y = 23\left(\frac{x}{5}\right)^{\frac{3}{2}}$

(D) $y = 23\left(\frac{x}{5}\right)^{\frac{2}{3}}$

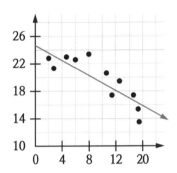

18. The scatterplot above shows a line of best fit for the data. For 11 data points on the scatterplot, how many data points does the line of best fit predicts a greater y-value than the actual y-value?

19. The value, in dollars, of a drone with camera t year after it has been purchased is modeled by the equation $g(t) = 53,000\,(0.78)^t$. By what percent does the value of the drone decrease each year?

(A) 12%

(B) 22%

(C) 78%

(D) 122%

20. In a survey, 2000 people were asked about a nation's economic status compared to the previous year. 736 people responded that the nation's economic status stayed the same. If the total population of the nation is 330 million peoples, what is the best estimate of the number of people who think the nation's economic status stayed the same?

(A) 121

(B) 209

(C) 12,140,000

(D) 121,000,000

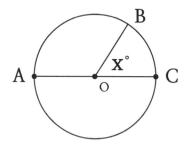

21. The circle above has center O, and $\overline{AC}$ is a diameter of the circle. If the radius of the circle is $\frac{5}{4}$ and the area of sector BOC is $\frac{5\pi}{32}$. What is the value of x ?

(A) 22.5

(B) 36

(C) 42

(D) 72

$$(x - 2) + (y + 3) = 43$$

$$(x - 2) - (y + 3) = 67$$

22. If (x, y) is the solution to the system of equations above, what is the value of $4(x - 2)$?

(A) 55

(B) 57

(C) 110

(D) 220

PRACTICE TEST 1
Advanced

Math
22 QUESTIONS

- The questions in this section address a number of important math skills.

- Use a calculator is permitted for all questions.

- Reference

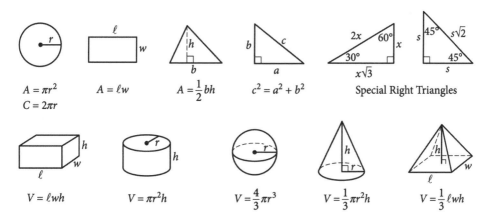

$A = \pi r^2$

$C = 2\pi r$

$A = \ell w$

$A = \frac{1}{2}bh$

$c^2 = a^2 + b^2$

Special Right Triangles

$V = \ell wh$

$V = \pi r^2 h$

$V = \frac{4}{3}\pi r^3$

$V = \frac{1}{3}\pi r^2 h$

$V = \frac{1}{3}\ell wh$

The number of degrees of arc in a circle is 360.

The number of radians of arc in a circle is 2π.

The sum of the measures in degrees of the angles of a triangle is 180.

For multiple-choice questions, solve each problem, choose the correct answer from the choices provided, and then circle your answer in this book. Circle only answer for each question. If you change your mind, completely erase the circle. You will not get credit for questions with more than one answer circled, or for questions with no answers circled.

For student-produced response questions, solve each problem and write your answer next to or under the question in the test book as described below.

- Once you've written your answer, circle it clearly. You will not receive credit for anything written outside the circle, or for any questions with more than one circled answer.

- **If you find more than one correct answer**, write and circle only one answer.

- Your answer can be up to 5 characters for a **positive** answer and up to 6 characters (Including the negative sign) for a **negative** answer, but no more.

- If your answer is a **fraction** that is too long (over 5 characters for positive, 6 characters for negative), write the decimal equivalent.

- If your answer is a **decimal** that is too long (over 5 characters for positive, 6 characters for negative), truncate it or round at the fourth digit.

- If your answer is a **mixed number** (such as $3\frac{1}{2}$), write it as an improper fraction $(7/2)$ or its decimal equivalent (3.5).

- Don't include **symbols** such as a percent sign, comma, or dollar sign in your circled answer.

Answer	Acceptable ways to enter answer	Unacceptable: will NOT receive credit
3.5	3.5 3.50 7/2	31/2 3 1/2
$\frac{2}{3}$	2/3 .6666 .6667 0.666 0.667	0.66 .66 0.67 .67
$-\frac{1}{3}$	$-\frac{1}{3}$ $-.3333$ -0.333	$-.33$ -0.33

1. A bag contains x apples, y oranges, and z pears. If the number of oranges in the bag is 40% of the sum of number of apples and pears, what is the probability of selecting a fruit that is not an orange?

    (A) 0.4

    (B) 0.714

    (C) 1.4

    (D) 2.5

2. A volume in gallons, of 53% saltwater solution will be mixed with a volume, in gallons, of a 20% saltwater solution to produce 40% saltwater solution. What volume, in gallons, of the 53% saltwater solution will be needed if 50 gallons of the 20% saltwater solutions is used?

$$y < \frac{1}{3}x + 4$$
$$y > -3x + 4$$

3. Which ordered pair (x, y) is a solution to the given system of inequalities in the xy-plane?

    (A) $(1, 0)$

    (B) $(6, 3)$

    (C) $(0, \ 4)$

    (D) $(3, -7)$

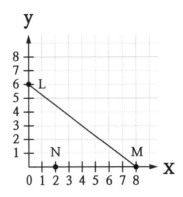

4. Line segment LM and point N are shown in the xy-plane. If line ℓ contain point N and is perpendicular to line segment LM, which of the following is an equation of line ℓ ?

(A) $-4x + 3y = -8$

(B) $-4x + 3y = 8$

(C) $3x + 4y = 6$

(D) $3x + 4y = -6$

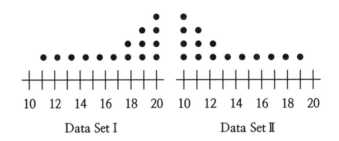

Data Set I   Data Set II

5. The dot plots show 16 data values from set I and 16 data values from set II. Which of the following statements best compares standard deviations of the values for each data set?

(A) The standard deviation of data set I is greater than the standard deviation of data set II.

(B) The standard deviation of data set I is smaller than the standard deviation of data set II.

(C) The standard deviation of data set I and data set II are the same.

(D) The standard deviation of data set I is twice the standard deviation of II.

6. The relationship between the estimated number of earthworms, y, in thousands, from a compost bin and the volume of the compost bin, x, in cubic meters can be modeled by the equation $y = 8.12x + 3.5$. Which of the following is the best interpretation of 8.12 in the context?

(A) The number of earthworms in the compost bin is estimated to increase by a factor of 8.12 for each additional cubic meter.

(B) The number of earthworms in the compost bin is estimated to decrease by a factor of 8.12 for each additional cubic meter.

(C) The number of earthworms in the compost bin is estimated to increase by 8.12 for each additional cubic meter.

(D) The number of earthworms in the compost bin is estimated to increase by 8120 for each additional cubic meter.

$$P(x) = -2x^2 + 28x + 98$$

7. The function shown models the number of dishes sold in a restaurant when \$$x$ per dish on $1 \leq x \leq 16$. What is the best estimate of the maximum number of dishes sold?

(A) 7

(B) 34

(C) 98

(D) 196

8. The lengths of the sides of a triangle are $9, 9$, and 5. What is the measure of the smallest angle in the triangle?

(A) 32.3°

(B) 33.4°

(C) 73.9°

(D) 115.4°

9. The height that a particular ball rebounds after it hits the ground is directly proportional to the height from which it falls. If the ball falls from an initial height of 270 meters, it hits the ground for the first time and rebounds to a height of 90 meters. How high does the ball rebound for the third time?

$$\frac{2}{\sqrt{5-x}} = \frac{1}{5}$$

10. What is the solution of the given equation?

11. A local newspaper includes a survey question about a current issue on the front page of the paper. Subscribers are invited to use social media to respond to the question. Concerning the opinions of the local population, which of the following is a plausible reason why the results of such surveys could be biased?

I. Newspaper subscribers are more likely to have strong opinions and respond than non-subscribers.

II. The opinions of newspaper subscribers are not necessarily representative of the local people.

III. Subscribers with access to social media are not necessarily representative of the population of a region.

(A) I only

(B) II only

(C) II and III only

(D) I, II and III

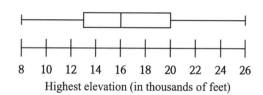

Highest elevation (in thousands of feet)

12. The highest elevation, in thousands of feet, was recorded for each of the 50 countries. The collected data contain no duplicate measurements and are summarized in the box plot. What is the interquartile range of the highest elevation in thousands of feet?

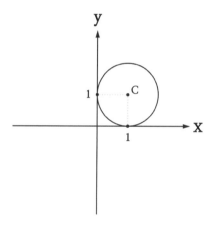

13. In the figure above, the circle has a center C at $(1,1)$. If the circle rolls along the positive x-axis for 2 revolution and then stops. Which of the following is an equation of the circle in its new position?

(A) $(x - (1 + 2\pi))^2 + (y - 1)^2 = 1$

(B) $(x - (1 + 4\pi))^2 + (y - 1)^2 = 1$

(C) $(x + (1 + 2\pi))^2 + (y - 1)^2 = 1$

(D) $(x + (1 + 4\pi))^2 + (y - 1)^2 = 1$

14. Which of the following data sets has the least standard deviation?

(A) $2, 2, 2, 10, 10, 10$

(B) $1, 2, 3, 4, 5, 6$

(C) $1, 1, 2, 2, 3, 3$

(D) $3, 3, 3, 3, 3, 3$

15. Which of the following expression is equivalent to $\frac{3-\sqrt{5}i}{3+\sqrt{5}i}$?

(A) $1 - \frac{3\sqrt{5}i}{2}$

(B) $1 - \frac{3\sqrt{5}i}{7}$

(C) $\frac{2}{7} - \frac{3\sqrt{5}i}{7}$

(D) $\frac{2}{7} + \frac{3\sqrt{5}i}{7}$

16. Given that $\cos x = -\frac{5}{12}$ and $\frac{\pi}{2} < x < \pi$, what is the value of $\sin x$?

(A) $-\frac{\sqrt{119}}{12}$

(B) $\frac{\sqrt{119}}{12}$

(C) $-\frac{\sqrt{119}}{5}$

(D) $\frac{\sqrt{119}}{5}$

Position	Sports Injuries		Total
	Yes	No	
Kicker	34	4	38
Linebacker	42	21	63
Quarterback	27	17	44
Receiver	47	8	55
Total	150	50	200

17. A survey asked 200 football players whether each one had ever gotten injured during games in a year. Each player in the list plays only one position. If a player is selected at random, what is the probability that the selected player will be Kicker or a Receiver, given that the player got injured in a year?

(A) $\frac{34}{200}$

(B) $\frac{34}{150}$

(C) $\frac{81}{200}$

(D) $\frac{81}{150}$

18. For which of the following values of x is $\cos x = -\sin x$?

(A) 0

(B) $\frac{\pi}{4}$

(C) $\frac{2\pi}{3}$

(D) $\frac{3\pi}{4}$

19.

$$f(x) = \begin{cases} 5 - \dfrac{1}{4}x^2 & for\ x < 3 \\ 3 & for\ x \geq 3 \end{cases}$$

If a function f is defined above, then $f(4) - f(-2)$ is

(A) -3

(B) -2

(C) -1

(D) 0

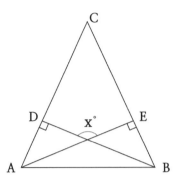

20. In the figure, $\overline{BD}$ is perpendicular to $\overline{AC}$ and $\overline{AE}$ is perpendicular to $\overline{BC}$. If the measure of $\angle A$ is $75°$, what is the value of x ?

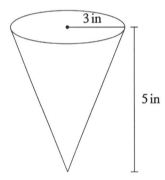

21. The figure above shows a conical drinking cup with a base of radius 3 inches and a height of 5 inches. What is the height h, in inches, of a conical drinking cup that is similar to the cup in the figure but has half the volume?

(A) 0.625

(B) 2.5

(C) 3.54

(D) 3.97

22. In the equation $x^2 + kx + 3 - k = 0$, k is a positive constant. If the equation has exactly one solution, what is the value of k?

Practice Test 1 Answers

\multicolumn{8}{c}{Practice Test 1-Module 1 ANSWERS}							
Number	Answer	Number	Answer	Number	Answer	Number	Answer
1	D	7	C	13	$\frac{2}{5} = 0.4 = .4$	19	D
2	D	8	3	14	C	20	B
3	B	9	$4.5 = \frac{9}{2}$	15	C	21	525
4	B	10	8	16	A	22	500
5	C	11	C	17	A		
6	D	12	B	18	C		

\multicolumn{8}{c}{Practice Test 1-Module 2-Easy ANSWERS}							
Number	Answer	Number	Answer	Number	Answer	Number	Answer
1	C	7	C	13	A	19	B
2	B	8	A	14	C	20	D
3	C	9	2	15	115	21	B
4	A	10	B	16	$\frac{4}{7} = 0.571 = .5714$	22	D
5	0	11	6	17	D		
6	A	12	C	18	5		

\multicolumn{8}{c}{Practice Test 1-Module 2-Advanced ANSWERS}							
Number	Answer	Number	Answer	Number	Answer	Number	Answer
1	B	7	D	13	B	19	C
2	76.92	8	A	14	D	20	150
3	B	9	10	15	C	21	D
4	A	10	−95	16	B	22	2
5	C	11	D	17	D		
6	D	12	7	18	D		

Explanation
Module 1

1. **(D)**

    Since Both triangle share A and right angles are congruent, two triangles are similar by AA similarity theorem. No additional information is needed.

2. **(D)**

    $$(\sqrt{x} - \sqrt{2y})^{\frac{2}{3}} = \sqrt[3]{x - 2\sqrt{2xy} + 2y}$$

3. **(B)**

   x stands for the number of Black turtle beans. $3x$ represents the total number of seeds from Black turtle beans.

4. **(B)**

   $18x + 6 = 9x + 6 + 8x$

   $18x + 6 = 17x + 6$

   Since slopes of two lines are different, they have one solution.

5. **(C)**

   1920 is 30 years since 1890.

   The actual annual percent growth rate is 0.24.

6. **(D)**

   $m = \frac{\Delta y}{\Delta x} = \frac{3}{4}$

   $y = \frac{3}{4}(x - 3) + 11 = \frac{3}{4}x + \frac{35}{4}$

7. **(C)**

   The age of the sediment is increasing by 4.5% every 100 meters below the top of the canyon.

8. **3**

   Let t be $(2x - 1)$.

   $t^2 - 10t + 25 = 0$

   $(t - 5)^2 = 0$

   $t = 5$

   $2x - 1 = 5$

   $x = 3$

9. **$4.5 = \frac{9}{2}$**

   By the axis of symmetry, $x = 2.25$, another time when the height of the volleyball will be 5 feet again is $2.25 + 2.25 = 4.5$ seconds.

10. **8**

    The location of median is $\frac{5+1}{2} = 3$.

    When we ordered the number of football games from the least to the greatest, the third number of football games is 8.

11. **(C)**

The equation of the given parabola is $y = 3(x+2)^2 + 4$. $h = 2$

12. **(B)**

The easiest possible combinations of number of T-shirts and pants she could buy are $(20, 0)$ or $(0, 10)$

20 T-shirts and 0 pants cost \$1000. 0 T-shirts and 10 pants cost \$1000.

Thus, Total amount of money she is spending is \$1000.

13. $\frac{2}{5} = 0.4 = .4$

$\tan B = \tan A = \frac{4}{10} = \frac{2}{5} = 0.4$

14. **(C)**

The function is an exponential function with a constant ratio, $b = 4$ and passes through $(1, 10)$.

$y = 10(4)^{x-1}$

15. **(C)**

$c(1 + \frac{55}{100}) = 1.55c$

16. **(A)**

$\frac{58 \times 1000}{3600} = 16.11$

17. **(A)**

$y = -2(0-1)(0+3)(0-4) = -24$

$(0, -24)$

18. **(C)**

I. In isosceles right triangle, base angles are congruent and each base angle is $45°$.

II. In isosceles right triangle, the ratio of side lengths is $1 : 1 : \sqrt{2}$.

19. **(D)**

$\cos 67° = \frac{5}{x}$.

$x = \frac{5}{\cos 67°} = 12.7965$

20. **(B)**

P(product of numbers is even)=1-P(product of numbers is odd).

$1 - \frac{3}{6} \times \frac{3}{6} = \frac{3}{4}$

21. **525**

The original mean is $\frac{4156}{6} = 692.6667$ The new mean is $\frac{837}{6} = 167.4$ The difference is approximately 525.

22. **500**

Density=$\frac{5.5}{0.011} = 500$

Module 2-Easy

1. **(C)**

when $x = -2, -8 + cy = 3$

$cy = 11$

$y = \frac{11}{c}$

$(-2, \frac{11}{c})$ is a solution.

2. **(B)**

The center of the new circle is $(1,0)$, but the radius becomes 2.

The equation of the new circle is $(x-1)^2 + y^2 = 4$.

3. **(C)**

$(2x+9)(1 - 10(2x+9)) = (2x+9)(-20x - 89) = -(20x + 89)(2x+9)$

4. **(A)**

$\frac{(x^2-9)(x-3)}{(x-3)^2} = \frac{(x-3)^2(x+3)}{(x-3)^2} = (x+3)$

5. **0**

The sum of two solutions is $-\frac{b}{a} = 0$

6. **(A)**

$V = 5 \cdot 6 \cdot 10 - \pi \cdot 4 \cdot 6 = 300 - 24\pi = 224.6$

7. **(C)**

$\frac{170}{100} \times 2000 = 3,400$

8. **(A)**

The initial value is $(0, 100)$ The decay factor is 0.4 every 5 increases in x.

$f(x) = 100(0.4)^{\frac{x}{5}}$

9. **2**

$\frac{120}{60} = 2$ rotations

10. **(B)**

Let x be the number of years since 2016.

$3,700 + 485x = 5640$.

$x = 4$

$2016 + 4 = 2020$

11. **6**

$5 : 2 = x : y$

$2x = 5y \rightarrow x = \frac{5y}{2}$

Area=$\frac{5y^2}{2} = 90$

$y = 6, x = 15$

The length of the shorter side is 6.

12. **(C)**

If a perpendicular line is drawn from E to BF, the height is 5 and the length of segment from the base of the height to F is 5 by the special right triangle ratio.

$BC = 11$ and $CD = 5$

Area of rectangle is 55.

13. **(A)**

The slope of line $3x = 4y + 2$ is $\frac{3}{4}$. The slope of perpendicular line to this is $-\frac{4}{3}$.

14. **(C)**

$d(v) = 7 \cdot v^2$

$d(3) = 7 \times 9 = 63$.

15. **115**

$m\angle ACB = 25° \ m\angle FAB = 90° + 25° = 110°$.

16. $\frac{4}{7} = 0.571 = .5714$

4 samples are alkaline.

$\frac{4}{7} = 0.571 = .5714$

17. **(D)**

$\left(\frac{x}{5}\right)^{\frac{2}{3}} = \frac{y}{23}$

$y = 23\left(\frac{x}{5}\right)^{\frac{2}{3}}$

18. **5**

Data points below the line of best fit are underestimated by the line. There are 5 data points below the line.

19. **(B)**

Percent decreases is $(1 - 0.78) \times 100 = 22\%$

20. **(D)**

$\frac{736}{2000} \times 330 \times 10^6 = 121,440,000$

21. **(B)**

$\pi\left(\frac{5}{4}\right)^2 \times \frac{x}{360} = \frac{5\pi}{32}$

$x = 36$

22. **(D)**

$2(x - 2) = 110$

$4(x - 2) = 220$

Module 2-Advanced

1. **(B)**

$y = 0.4(x + z)$

P(not an orange)$=\frac{x+z}{x+y+z} = \frac{x+z}{1.4(x+z)} = \frac{1}{1.4} = 0.71428.$

2. **76.92**

$0.53x + 0.2(50) = 0.4(x + 50)$

$x = 76.923 \approx 76.92$

3. **(B)**

When $x = 6, y < 5$ and $y > -14$.

$(6, 3)$ is in the solution set.

4. **(A)**

$y = \frac{4}{3}(x - 2) = \frac{4}{3}x - \frac{8}{3}.$

$-4x + 3y = -8.$

5. **(C)**

The standard deviation is calculated by the average distance of the numbers in each list from the mean. The data sets I and II have the same standard deviation because the distributions are symmetrical, and the absolute distance of data points from the mean is the same.

6. **(D)**

$m = \frac{\Delta y}{\Delta x} = 8.12\text{thousands} = 8,120$

The increase in the number of earthworms is 8,120 for each additional cubic meter in the volume of the compost bin.

7. **(D)**

The maximum number of dishes sold occurs at the vertex.

$h = -\frac{b}{2a} = \frac{28}{4} = 7$

$k = P(7) = 196$

The maximum number of dishes sold is 196.

8. **(A)**

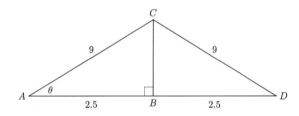

$\cos\theta = \frac{2.5}{9} \quad \theta = 73.87° \quad m\angle ACB = 16.127° \quad m\angle ACD = 32.255°$

9. **10**

$y = 270(\frac{1}{3})^x$

$y = 270(\frac{1}{3})^3 = 10$

10. **−95**

$10 = \sqrt{5 - x} \quad 100 = 5 - x \quad x = -95$

11. **(D)**

The sampling method is biased because it does not represent all the local people. Newspaper subscribers are more likely to have strong opinions than non-subscribers. On top of that, those who have difficulty accessing social media are not included in the sample.

I, II and III are true.

12. **7**

$$IQR = Q_3 - Q_1 = 20 - 13 = 7$$

13. **(B)**

It moves 4π along the x-axis. The new center is $(1 + 4\pi, 1)$.

The equation of circle is $(x - (1 + 4\pi))^2 + (y - 1)^2 = 1$

14. **(D)**

(D) All the data values are equal, so the standard deviation is 0.

15. **(C)**

$$\frac{(3 - \sqrt{5}i)^2}{(3 + \sqrt{5}i)(3 - \sqrt{5}i)} = \frac{4 - 6\sqrt{5}i}{14} = \frac{2}{7} - \frac{3\sqrt{5}}{7}i$$

16. **(B)**

Since the angle x in the second quadrant, $\sin x$ is positive. Opposite side of the angle is $\sqrt{12^2 - 5^2} = \sqrt{119}$

$\sin x = \frac{\sqrt{119}}{12}$

17. **(D)**

$$P(\text{Kicker or Receiver} \mid \text{Sports Injuries}) = \frac{34 + 47}{150} = \frac{81}{150}$$

18. **(D)**

$\frac{\sin x}{\cos x} = -1$

$\tan x = -1$

$x = \frac{3\pi}{4}$

19. **(C)**

$f(4) = 3, f(-2) = 4$

$f(4) - f(-2) = -1$

20. **150**

$\triangle ADB \cong \triangle BEA$ by AAS.

$m\angle A = m\angle B = 75°$

$m\angle EAB = m\angle DBA = 15°$

$x° = 180 - 15 - 15 = 150°$

21. **(D)**

Ratio of volumes is $1 : \frac{1}{2}$

Ratio of lengths is $1 : \frac{1}{\sqrt[3]{2}}$

$1 : \frac{1}{\sqrt[3]{2}} = 5 : h$

$h = \frac{5}{\sqrt[3]{2}} = 3.9685$

22. **2**

$D = k^2 - 4(3 - k) = 0$

$k^2 + 4k - 12 = (k + 6)(k - 2) = 0$

$k = 2, -6$

Since k is a positive constant, $k = 2$

PRACTICE TEST 2

Math

22 QUESTIONS

- The questions in this section address a number of important math skills.

- Use a calculator is permitted for all questions.

- Reference

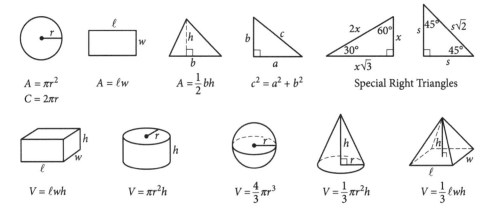

$A = \pi r^2$
$C = 2\pi r$

$A = \ell w$

$A = \frac{1}{2} bh$

$c^2 = a^2 + b^2$

Special Right Triangles

$V = \ell w h$

$V = \pi r^2 h$

$V = \frac{4}{3}\pi r^3$

$V = \frac{1}{3}\pi r^2 h$

$V = \frac{1}{3}\ell w h$

The number of degrees of arc in a circle is 360.

The number of radians of arc in a circle is 2π.

The sum of the measures in degrees of the angles of a triangle is 180.

For multiple-choice questions, solve each problem, choose the correct answer from the choices provided, and then circle your answer in this book. Circle only answer for each question. If you change your mind, completely erase the circle. You will not get credit for questions with more than one answer circled, or for questions with no answers circled.

For student-produced response questions, solve each problem and write your answer next to or under the question in the test book as described below.

- Once you've written your answer, circle it clearly. You will not receive credit for anything written outside the circle, or for any questions with more than one circled answer.

- **If you find more than one correct answer**, write and circle only one answer.

- Your answer can be up to 5 characters for a **positive** answer and up to 6 characters (Including the negative sign) for a **negative** answer, but no more.

- If your answer is a **fraction** that is too long (over 5 characters for positive, 6 characters for negative), write the decimal equivalent.

- If your answer is a **decimal** that is too long (over 5 characters for positive, 6 characters for negative), truncate it or round at the fourth digit.

- If your answer is a **mixed number** (such as $3\frac{1}{2}$), write it as an improper fraction (7/2) or its decimal equivalent (3.5).

- Don't include **symbols** such as a percent sign, comma, or dollar sign in your circled answer.

Answer	Acceptable ways to enter answer	Unacceptable: will NOT receive credit
3.5	3.5 3.50 7/2	31/2 3 1/2
$\frac{2}{3}$	2/3 .6666 .6667 0.666 0.667	0.66 .66 0.67 .67
$-\frac{1}{3}$	$-\frac{1}{3}$ $-.3333$ -0.333	$-.33$ -0.33

1. Which polynomial is equivalent to $(x^2 - 7)(12x^3 + 4)$?

   (A) $12x^5 - 28$

   (B) $12x^5 - 84x^3 + 4x^2 - 28$

   (C) $12x^6 - 84x^3 + 4x^2 - 28$

   (D) $12x^6 + 4x^3 - 84x^2 - 28$

$$3x + y = 12$$

2. One of the two linear equations in a system is given. The system has no solution. Which equation could be the second equation in the system?

   (A) $y = -3(x + 3)$

   (B) $y = -3(x - 4)$

   (C) $y = 3(x + 3)$

   (D) $y = 4(x + 3)$

3. For acute angles A and B, $\sin A = \cos B$. The measure of angle A is $46°$. What is the measure of angle B?

   (A) $34°$

   (B) $44°$

   (C) $46°$

   (D) $134°$

4. Jennifer needs to water her vegetable garden bed for an average of 80 gallons. She was recommended to water the garden bed at an average of 20 gallons per 32 square feet. How big is her vegetable garden bed, in square feet?

Class	Drawing Type			Total
	Figurative	Realistic	Hyperrealistic Drawing	
A	4	8	6	18
B	7	5	10	22
Total	11	13	16	40

5. The table shows the number of drawings by drawing type exhibited in two different art classes. If a hyperrealistic drawing is selected at random, which of the following is closest to the probability that the selected drawing is from class A?

(A) 0.10

(B) 0.22

(C) 0.38

(D) 0.62

6. The expression $x^2 + 6x - 7$ can be written in the form $a^2 - b^2$, where a is an expression and b is a constant. Which of the following gives an expression for a and the value of b ?

(A) $a = x + 3,\ b = 16$

(B) $a = x + 3,\ b = 4$

(C) $a = x + 6,\ b = 16$

(D) $a = x + 6,\ b = 4$

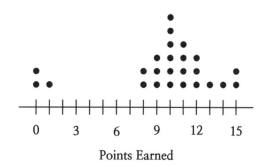

Points Earned

7. A group of students played a game in which they earned points for answering questions correctly. What is the mean of the 25 points represented in the dot plot above?

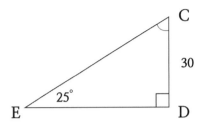

8. Triangle CDE(shown) is similar to triangle PQR(not shown). For these triangles, $\angle D$ and $\angle E$ correspond to $\angle Q$ and $\angle R$, respectively, and $CE = 3PR$. Which of the following statements is(are) true?

I. The measure of $\angle R$ is $75°$.

II. $PQ = 90$

(A) I only

(B) II only

(C) I and II

(D) Neither I nor II

9. People enter a line for an escalator, and there are already 20 people in the line. The number of people who enter a line for an escalator is modeled by the function $N(t) = 20 - 0.5t$, where t is measured in second. Which of the following is the best interpretation of $N(40)$ in this context?

(A) The number of people in the line for an escalator 40 seconds after the first time the number of people who entered the line has counted

(B) The number of people in the line for an escalator 0 seconds after the first time the number of people who entered the line has counted

(C) The number of seconds since the beginning of the escalator when there are 40 people in the line.

(D) The number of seconds since the beginning of the escalator when there are 0 people in the line.

10. What is the graph of $y = -4x + 5$?

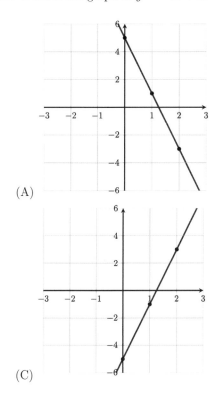

(A)

(C)

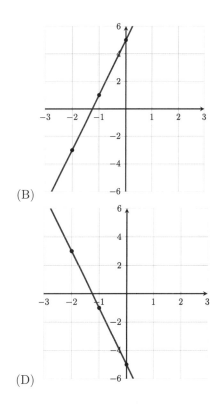

(B)

(D)

$$A_1 = 5000 \, (0.876)^m$$

$$A_2 = 5000 - 300.5m$$

11. An exponential and a linear equation estimate the area covered by water, in square miles, in South Carolina, m months after January 1 in 2021, for $0 \leq m \leq 6$. The area covered by water, in square miles, estimated by the linear equation is how much greater than the area covered by water estimated by the exponential equation on April 1,2021?

(A) 0.737

(B) 7.37

(C) 73.7

(D) 737

12. What is the y-intercept of the graph of $y = 5^x + 10$ in the xy-plane?

13. The measure of angle α is $\frac{\pi}{12}$ radians greater than the measure of β. How much greater is the measure of angle α than the measure of angle β, in degrees? (Disregard the degree symbol when entering your answer.)

14. If $m\%$ more than 300 is 500, what is $m\%$ of 300?

    (A) 120

    (B) 180

    (C) 200

    (D) 500

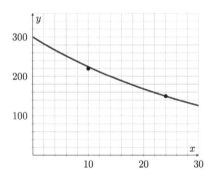

15. The graph shown models the value, in dollars, of a cryptocurrency over time, in years, after an original amount is deposited. According to the graph, how many years does it take for the value of the cryptocurrency decrease to one-half of its original amount?

    (A) 12

    (B) 14

    (C) 22

    (D) 24

$$4, \ 4, \ 4, \ 5, \ 5, \ 7, \ 7, \ 8, \ 9, \ 9, \ 10$$

16. The list shows the eleven ratings of the participants about their satisfaction with the air quality at work. What is the median number of ratings from eleven participants?

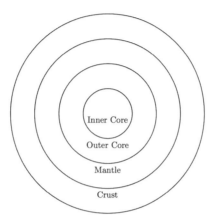

17. A cross-section of the planet Earth is shown. Each of the 4 layers is spherical. The radius of Earth is approximately 6371 kilometers, and the radius of the outer core is approximately 3540 kilometers. Which of the following is closest to the total volume, in cubic kilometers, of the mantle and crust combined?

(A) 88,100,000

(B) 95,000,000,000

(C) 504,800,000,000

(D) 897,000,000,000

18. The ratio of students and teachers in a tutoring seminar is 12 to 1. If there are x students at the seminar, what expression represents the number of teachers at the seminar in terms of x ?

(A) $12x$

(B) $x + 12$

(C) $\frac{x}{12}$

(D) $\frac{12}{x}$

$$x^2 + y^2 + 6x - 5y = -\frac{25}{4}$$

19. The equation of a circle in the xy-plane is shown. What is the y-coordinate of the center?

$$4x + \frac{2}{3}y = 7$$

$$5x - \frac{1}{3}y = \frac{31}{2}$$

20. The solution to the given system of equations is (x, y). What is the value of $x - y$?

21.

$$4x^4 - 28x^2y + 48y^2$$

Which of the following is the factor of the given expression?

(A) $x - 3y$

(B) $x^2 - 3y$

(C) $4x^2 + 9y$

(D) $x^2 + 4y$

$$f(x) = -\frac{1}{16}x^2 + 250$$

22. The given equation defines the function f. For what value of x does $f(x)$ reach its maximum?

PRACTICE TEST 2
Easy

Math
22 QUESTIONS

- The questions in this section address a number of important math skills.

- Use a calculator is permitted for all questions.

- Reference

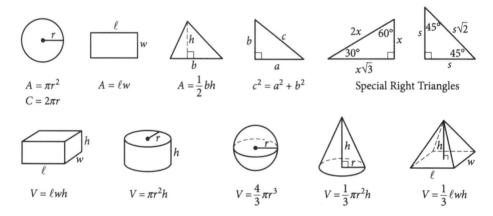

$A = \pi r^2$
$C = 2\pi r$

$A = \ell w$

$A = \frac{1}{2}bh$

$c^2 = a^2 + b^2$

Special Right Triangles

$V = \ell wh$

$V = \pi r^2 h$

$V = \frac{4}{3}\pi r^3$

$V = \frac{1}{3}\pi r^2 h$

$V = \frac{1}{3}\ell wh$

The number of degrees of arc in a circle is 360.

The number of radians of arc in a circle is 2π.

The sum of the measures in degrees of the angles of a triangle is 180.

For multiple-choice questions, solve each problem,choose the correct answer from the choices provided, and then circle your answer in this book. Circle only answer for each question. If you change your mind, completely erase the circle. You will not get credit for questions with more than one answer circled, or for questions with no answers circled.

For student-produced response questions, solve each problem and write your answer next to or under the question in the test book as described below.

- Once you've written your answer, circle it clearly. You will not receive credit for anything written outside the circle, or for any questions with more than one circled answer.

- **If you find more than one correct answer**, write and circle only one answer.

- Your answer can be up to 5 characters for a **positive** answer and up to 6 characters (Including the negative sign) for a **negative** answer, but no more.

- If your answer is a **fraction** that is too long (over 5 characters for positive, 6 characters for negative), write the decimal equivalent.

- If your answer is a **decimal** that is too long (over 5 characters for positive, 6 characters for negative), truncate it or round at the fourth digit.

- If your answer is a **mixed number** (such as $3\frac{1}{2}$), write it as an improper fraction (7/2) or its decimal equivalent (3.5).

- Don't include **symbols** such as a percent sign, comma, or dollar sign in your circled answer.

Answer	Acceptable ways to enter answer	Unacceptable: will NOT receive credit
3.5	3.5 3.50 7/2	31/2 31/2
$\frac{2}{3}$	2/3 .6666 .6667 0.666 0.667	0.66 .66 0.67 .67
$-\frac{1}{3}$	$-\frac{1}{3}$ $-.3333$ -0.333	$-.33$ -0.33

1. How many solutions does the equation $3x - 10 = x + 2(x - 5)$ have?

   (A) Zero

   (B) Exactly one solution

   (C) Exactly two solutions

   (D) Infinitely many solutions

2. The graph of a line in the xy-plane passes through the points with coordinates $(6, 4)$ and $(10, 2)$. The line crosses the y-axis at the point with coordinate $(0, b)$. What is the value of b?

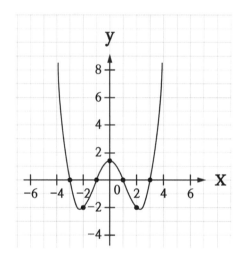

3. The graph of the function f is shown. What is the possible factor of $y = f(x)$?

   (A) $x(x + 1)$

   (B) $(x - 1)(x + 1)$

   (C) $(x - 2)(x + 1)$

   (D) $(x + 3)(x + 2)$

4. Triangles ABC and PQR each have a corresponding angle measuring 35° and a corresponding angle measuring 50°. Which additional piece of information is sufficient to determine whether triangle ABC is congruent to triangle PQR ?

(A) The degree of one pair corresponding angles from the two triangles

(B) The length of one pair corresponding sides from the two triangles

(C) The perimeter of triangle ABC

(D) No additional piece of information is necessary to determine whether the two triangles are congruent.

5. When the temperature of seawater is 10 degrees Celsius, sound travels through the seawater at a constant speed of about 1,490 meters per second. At the same temperature, approximately how far, in meters, would sound travel through the seawater in 1 minute?

(A) 89,400

(B) 14,900

(C) 22,500

(D) 25

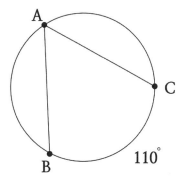

6. When the measure of arc BC is 110°, what is the measure, in degrees, of $\angle BAC$?

$$|2x + 1| = 49$$

7. What are all possible solutions to the given equation?

    (A) -25 and 24

    (B) -24 and 25

    (C) 3

    (D) 24

$$y \leq \frac{1}{2}x + 3$$
$$y > -4x - 5$$

8. Which ordered pair (x, y) is NOT a solution to the given system of inequalities in the xy-plane?

    (A) $(0, 3)$

    (B) $(-1, 0)$

    (C) $(1, -10)$

    (D) $(2, 0)$

$$y = \frac{1}{2}x - 18$$

9. One of the equations in a system of two linear equations is given. The system has one solution. Which equation could be the second equation in the system?

    (A) $y = \frac{1}{2}x$

    (B) $y = \frac{1}{2}x - 18$

    (C) $y = -2x$

    (D) $y = \frac{1}{2}x + 18$

10. An object was launched into the air. The equation $h = -16(t-7)^2 + 784$ represents this situation, where h is the height of the object above the ground in meters, t seconds after it was launched. Based on the model, what was the maximum height, in meters?

(A) 784

(B) 14

(C) 7

(D) 0

$$\frac{x^{a^2}}{x^{b^2}} = x^{36}$$

11. If the equation shown is true for $x > 1$ and $a + b = 2$, what is the value of $a - b$?

(A) 6

(B) 14

(C) 16

(D) 18

12. Cube A has a side length of x meters. Each side length of cube B is 8 times as long as each side length of cube A. What is the volume of cube B, in meters, in terms of x ?

(A) $2x^3$

(B) $8x^3$

(C) $64x^3$

(D) $512x^3$

$$R = \frac{9}{5}(C + 273.15)$$

13. The equation above expresses the temperate R, in degrees Rankine, in terms of the temperature C, in degrees Celsius. Which of the following expresses the temperature in degrees Celsius in terms of the temperature in degrees Rankine?

(A) $C = \frac{5}{9}(R - 491.67)$

(B) $C = \frac{5}{9}R + 273.15$

(C) $C = \frac{9}{5}(R - 491.67)$

(D) $C = \frac{9}{5}R + 273.15$

x	−1	1	3
f(x)	2	18	162

14. For the exponential function f, the table shows some values of x and the corresponding values of $f(x)$. The function can be written in the form $f(x) = a \cdot b^x$, where a and b are constants. What is the value of b ?

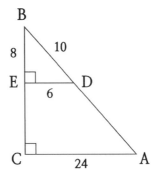

15. In the figure, triangle ABC and DBE are right triangles. What is $\cos B$?

16. Which of the following represents the positive number x increased by 150% ?

(A) $0.05x$

(B) $1.50x$

(C) $2.50x$

(D) $150x$

17. The table shows the distribution by breed and sex of dogs that were sent to animal adoption shelter during the month of July.

		Sex	
		Male	Female
Breed	American Staffordshire Terrier	6	2
	Alaskan malamute	4	8

If a male dog that was sent to the shelter during the month of July is selected at random, what is the probability that the dog breed is American Staffordshire Terrier?

(A) 0.33

(B) 0.40

(C) 0.60

(D) 0.75

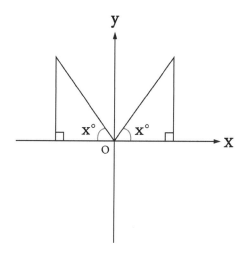

18. In the coordinate plane, $\sin x° = \frac{5}{8}$. What is the value of $\cos(180° - x°)$?

(A) $-\frac{\sqrt{39}}{8}$

(B) $\frac{\sqrt{39}}{8}$

(C) $-\frac{5}{8}$

(D) $\frac{5}{8}$

State	Number of road bridges in 2020
Massachusetts	5,245
New Jersey	6,798
Oregon	8,235
Ohio	27,151
Alabama	16,164
Wiscosin	14,307
Texas	55,175

19. The table shows the number of road bridges in 2020 for 7 US states. Based on the given information, what is the median number of road bridges among 7 US states?

(A) 27,151

(B) 16,164

(C) 14,307

(D) 19,011

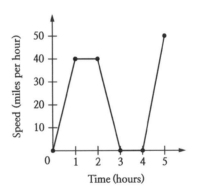

20. The graph above shows Lucas's speed during a road trip. He stopped at a rest stop during his trip to take a nap. Based on the graph, how many miles have he traveled before he stopped for his nap?

(A) 40

(B) 50

(C) 60

(D) 80

$$x + y = 1$$

$$3x + 3y = 1$$

21. How many solutions does the given system of equations have?

    (A) Zero

    (B) Exactly one

    (C) Exactly two

    (D) Infinitely many

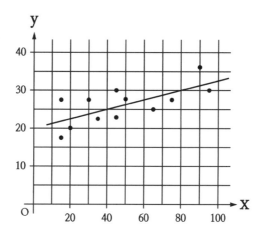

22. The scatterplot shows twelve data points and a line of best fit for the data. What is the slope of the line?

    (A) $\frac{1}{8}$

    (B) $\frac{1}{4}$

    (C) $\frac{1}{3}$

    (D) $\frac{1}{2}$

PRACTICE TEST 2
Advanced

Math
22 QUESTIONS

- The questions in this section address a number of important math skills.

- Use a calculator is permitted for all questions.

- Reference

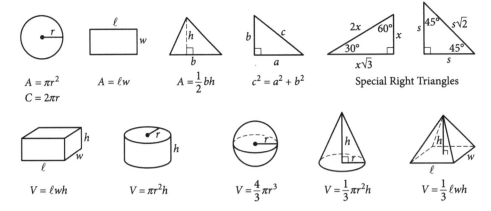

$A = \pi r^2$
$C = 2\pi r$

$A = \ell w$

$A = \frac{1}{2}bh$

$c^2 = a^2 + b^2$

Special Right Triangles

$V = \ell wh$

$V = \pi r^2 h$

$V = \frac{4}{3}\pi r^3$

$V = \frac{1}{3}\pi r^2 h$

$V = \frac{1}{3}\ell wh$

The number of degrees of arc in a circle is 360.

The number of radians of arc in a circle is 2π.

The sum of the measures in degrees of the angles of a triangle is 180.

For **multiple-choice questions**, solve each problem,choose the correct answer from the choices provided, and then circle your answer in this book. Circle only answer for each question. If you change your mind, completely erase the circle. You will not get credit for questions with more than one answer circled, or for questions with no answers circled.

For **student-produced response questions**, solve each problem and write your answer next to or under the question in the test book as described below.

- Once you've written your answer, circle it clearly. You will not receive credit for anything written outside the circle, or for any questions with more than one circled answer.

- **If you find more than one correct answer**, write and circle only one answer.

- Your answer can be up to 5 characters for a **positive** answer and up to 6 characters (Including the negative sign) for a **negative** answer, but no more.

- If your answer is a **fraction** that is too long (over 5 characters for positive, 6 characters for negative), write the decimal equivalent.

- If your answer is a **decimal** that is too long (over 5 characters for positive, 6 characters for negative), truncate it or round at the fourth digit.

- If your answer is a **mixed number** (such as $3\frac{1}{2}$), write it as an improper fraction (7/2) or its decimal equivalent (3.5).

- Don't include **symbols** such as a percent sign, comma, or dollar sign in your circled answer.

Answer	Acceptable ways to enter answer	Unacceptable: will NOT receive credit
3.5	3.5 3.50 7/2	31/2 31/2
$\frac{2}{3}$	2/3 .6666 .6667 0.666 0.667	0.66 .66 0.67 .67
$-\frac{1}{3}$	$-\frac{1}{3}$ $-.3333$ -0.333	$-.33$ -0.33

1. $f(x) = 2x + 1$ and $g(x) = x^2 + 2x - 1$, which of the following is equal to $g(f(x))$?

   (A) $4x^2 + 4x - 1$

   (B) $4x^2 + 8x + 2$

   (C) $4x^2 + 4x + 2$

   (D) $2x^3 + 5x^2 - 1$

2. A right rectangular prism has length x, width w, and height h. If $h = 5x$ and $w = 4x$, what is the volume of the prism in terms of x?

   (A) $16x^3$

   (B) $20x^3$

   (C) $25x^3$

   (D) $80x^3$

3. A quiz game contains two types of 40 questions, some worth 3 points, and others worth 8 points. If a student achieved a total score of 150 points, what is the difference in the number of 3 points questions and 8 points questions?

4. If $x > 0, y = \frac{1}{x} + \frac{2}{x}$. Which of the following is the reciprocal of y?

   (A) $\frac{x}{3}$

   (B) $\frac{3}{x}$

   (C) $\frac{2x}{3}$

   (D) $\frac{3}{2x}$

$$a, 8, 18, b, 30, 32, 32, 40$$

5. The eight numbers above, given in ascending order, have a mean of 24 and a median of 28. Find the value of a.

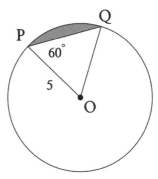

6. In the figure above, the circle has center O and the radius of the circle is 5. What is the area of the shaded region?

(A) $\frac{25\pi}{6}$

(B) $\frac{1}{12}\left(50\pi - 75\sqrt{3}\right)$

(C) $\frac{1}{12}\left(75\sqrt{3} - 25\pi\right)$

(D) $\frac{1}{12}\left(50\pi - 150\sqrt{3}\right)$

7. A ball is kicked from a ground. The equation $h = -4.9t^2 + 15t$ represent this situation, where h is the height of the ball above ground, in meters, t seconds after it is kicked. At what time, in second will be the ball reach the maximum height?

(A) 1

(B) 1.53

(C) 3.06

(D) 5

8. During an experiment, the number of phytoplankton in a population is growing by roughly 1.5 percent per day. There was 600 phytoplankton at the start of the experiment. In how many days will the population first exceed 1000?

(A) 32

(B) 33

(C) 34

(D) 35

9. Salesperson considers offers from two different companies. Company X gives $1,525 plus a 10% commission on his sales each week. Company Y suggests $520 plus a 25% commission on his sales each week. Which inequality models the amount of sales each week, d dollars for which Company X gives a better offer than Company Y?

(A) $d < 6,700$

(B) $d > 6,700$

(C) $d > 2,871.4$

(D) $d < 2,871.4$

10. A store owner bought 20 dozen eggs at a cost of $3 per dozen. She sold all of the eggs in one day for $0.50 each. What was her profit?

11. Line m has a slope of $-\frac{1}{10}$ and passes through $(\frac{1}{2}, 4)$. Which of the following defines line m?

(A) $2x + 20y = 81$

(B) $2x + 20y = 79$

(C) $2x - 20y = 81$

(D) $2x - 20y = 79$

12. Square A has an area 16 square inches. Square B has an area that is 128 square inches greater than that of Square A. Which of the following lengths, in inches, is closest to the side length of Square B?

(A) 10.6

(B) 11.3

(C) 12.0

(D) 15.3

13. A quality control inspector must check the mean weight of bags of chips to verify whether a machine works correctly. The inspector will randomly select a sample of 100 bags and weigh each of bag. The mean weight of packages of chips from the sample is 9.8 ounces. Assume that the margin of error of weight is 0.3 ounces. What is the possible mean weight of bags of chips in the factory?

(A) 9.2

(B) 9.4

(C) 9.6

(D) 10.2

14. A car travels east at an average speed of 80 miles per hour for 1 hour and then north at an average speed of 50 miles per hour for 2 hours. What is the shortest distance from the starting point to the end of the trip?

(A) 94.34

(B) 128.06

(C) 130

(D) 180

15. Mia, Justin, and Conan work together in the fundraising booth. They decided to take turns watching the booth and receive the money received in the same proportion as the time each spent watching the booth. Justin watched 5 hours longer than Mia, and Conan watched 3 hours less than Justin. If the booth was open for 12 hours, what percent of the money did Mia receive?

(A) 13.9%

(B) 30.6%

(C) 33.3%

(D) 55.6%

16. For a data set that consists of 10 values, the median is much smaller than the mean. Which of the following could explain why the median is much smaller than the mean?

(A) The distribution of values from the data set is symmetric about the median.

(B) The data set contains a value that is extremely large relative to other values in the data set.

(C) The data set contains a value that is extremely small relative to other values in the data set.

(D) Each value in the data set is the same.

17. In the xy-plane, the equation of parabola $f(x) = -x^2 + bx + c$, where b and c are constants, has a vertex $(3, -14)$. Which of the following is $b + c$?

(A) -23

(B) -17

(C) -13

(D) -7

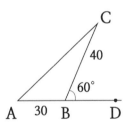

18. In the figure shown above, $AB = 30$ inches and $BC = 40$ inches and the measure of $\angle CBD = 60°$. To nearest inches, what is the length of AC ?

(A) 36

(B) 46

(C) 61

(D) 78

n	f(n)
1	3
2	9
3	19
4	33
5	51

19. The table above gives selected values for a function f. Which of the following could be an expression for $f(n)$?

(A) $6n - 3$

(B) $4^n - 1$

(C) $\frac{3n^2 + 3n}{2}$

(D) $2n^2 + 1$

$$2^{x+1} + 2^{x-1}$$

20. Which of the following is equivalent to the expression above?

(A) 2^{2x}

(B) $3 \cdot 2^{x-1}$

(C) $5 \cdot 2^{x-1}$

(D) $5 \cdot 2^{x}$

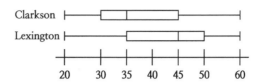

21. The boxplots above summarize age groups in music festivals in two cities, Clarkson and Lexington. Based on the boxplots, which of the following statements about age groups in two cities CAN NOT be true?

(A) The range of age groups in Clarkson is equal to the range of age groups in Lexington.

(B) The interquartile range of age groups in Clarkson is equal to the interquartile range of age groups in Lexington.

(C) About 75% of the audience at the festival in Lexington are older than or equal to about 50% of audience at the festival in Clarkson.

(D) Lexington and Clarkson have the same number of audiences.

$$x^2 - 2x + y^2 - 4y - 4 = 0$$

$$y = x - 2$$

22. If (x, y) is the solution to the system of equations above, what is the possible value of x ?

Practice Test 2 Answers

Practice Test 2-Module 1 ANSWERS							
Number	Answer	Number	Answer	Number	Answer	Number	Answer
1	B	7	9.64	13	15	19	$\frac{5}{2} = 2.5$
2	A	8	D	14	C	20	$\frac{17}{2} = 8.5$
3	B	9	A	15	D	21	B
4	128	10	A	16	7	22	0
5	C	11	D	17	D		
6	B	12	11	18	C		

Practice Test 2-Module 2-Easy ANSWERS							
Number	Answer	Number	Answer	Number	Answer	Number	Answer
1	D	7	A	13	A	19	C
2	7	8	C	14	3	20	D
3	B	9	C	15	$\frac{4}{5} = .8 = 0.8$	21	A
4	B	10	A	16	C	22	A
5	A	11	D	17	C		
6	55	12	D	18	A		

Practice Test 2-Module 2-Advanced ANSWERS							
Number	Answer	Number	Answer	Number	Answer	Number	Answer
1	B	7	B	13	C	19	D
2	B	8	D	14	B	20	C
3	28	9	A	15	A	21	D
4	A	10	60	16	B	22	1, 4
5	6	11	A	17	B		
6	B	12	C	18	C		

Explanation
Module 1

1. **(B)**

   $12x^5 + 4x^2 - 84x^3 - 28$

2. **(A)**

   The first linear equation is $y = -3x + 12$.

   (A) $y = -3x - 9$. This has the same slope but a different y-intercept as the first one. Thus, the system has no solution.

3. **(B)**

$A + B = 90°$

$B = 90 - 46 = 44$

4. **128**

$20 : 32 = 80 : A$

$A = 128$

5. **(C)**

$P(\text{class A} \mid \text{hyper realistic drawing}) = \frac{6}{16} = 0.375$

6. **(B)**

$x^2 + 6x - 7 = (x^2 + 6x + 9) - 7 - 9 = (x+3)^2 - 16 = (x+3)^2 - 4^2$

$a = x + 3, b = 4$

7. **9.64**

The mean is $\frac{1+16+27+60+44+36+13+14+30}{25} = 9.64$

8. **(D)**

I. Corresponding angles are congruent in two similar triangles. This statement is false.

II. $CD = 3$. $CD = 30, PQ = 10$ This statement is false.

9. **(A)**

$N(40)$ represents the number of people who enter a line for an escalator after 40 seconds.

10. **(A)**

$y = -4x + 5$ has y-intercept at $(0, 5)$ and passes through $(2, -3)$.

11. **(D)**

$5,000 - 300.5(3) - 5,000(0.876)^3 = 737.39$

12. **11**

The y-intercept is $(0, 11)$.

13. **15**

$\frac{\pi}{12} = \frac{180°}{12} = 15$

14. **(C)**

$300(1 + \frac{m}{100}) = 500$

$m = 66.6\%$

$0.666 \times 300 = 200$

15. **(D)**

The initial value of a cryptocurrency is $300. It becomes $150 about 24 years later.

16. **7**

The location of median is $\frac{12}{2} = 6$

The 6th number in the ordered list is 7.

17. **(D)**

$V = \frac{4\pi}{3}(6371^3 - 3540^3) = 8.9738 \times 10^{11}$

18. **(C)**

$12 : 1 = x : T$

$x = 12T$

$T = \frac{x}{12}$

19. $\frac{5}{2}$

$(x^2 + 6x + 9) + (y^2 - 5y + \frac{25}{4}) = -\frac{25}{4} + \frac{25}{4} + 9$

$(x + 3)^2 + (y - \frac{5}{2})^2 = 9$

20. $\frac{17}{2} = \mathbf{8.5}$

$x - y = \frac{31}{2} - 7 = \frac{17}{2} = 8.5$

21. **(B)**

$4x^4 - 28x^2y + 48y^2 = (4x^2 - 16y)(x^2 - 3y)$

22. **0**

The vertex is $(0, 250)$.

The maximum value occurs at $x = 0$.

Module 2-Easy

1. **(D)**

   $3x - 10 = 3x - 10$

   The slopes are equal and y-intercepts are equal.

   The system has infinitely many solutions.

2. **7**

   $m = -\frac{1}{2}$

   $y = -\frac{1}{2}(x - 10) + 2 = -\frac{1}{2}x + 7.$

   $(0, 7)$

   $b = 7$

3. **(B)**

   The factors of f are $(x + 3), (x + 1), (x - 1)$ and $(x - 3)$.

   (B) $(x + 1)(x - 1)$ is the factor of f.

4. **(B)**

   IF one pair of corresponding sides are congruent, we can prove that two triangles are congruent by AAS or ASA.

5. **(A)**

   $1,490 \times 60 = 89,400$

6. **55**

   $m\angle BAC = \frac{110°}{2} = 55°$

7. **(A)**

   $2x + 1 = \pm 49$

   $2x = \pm 49 - 1$

   $x = -25, 24$

8. **(C)**

   When $x = 1$, $y \le 2.5$ and $y > -9$.

   $(1, -10)$ is NOT in the solution set.

9. **(C)**

The second linear equation should have a different slope, which is not equal to $\frac{1}{2}$.

10. **(A)**

The vertex is $(7, 784)$. The maximum height is 784.

11. **(D)**

$x^{a^2 - b^2} = x^{36}$ $a^2 - b^2 = (a+b)(a-b) = 36$

$a - b = 18$

12. **(D)**

The length of cube B is $8x$. The volume of cube B is $512x^3$.

13. **(A)**

$\frac{5}{9}R = C + 273.15$

$C = \frac{5}{9}R - 273.15 = \frac{5}{9}(R - 491.67)$

14. **3**

$ab^{-1} = 2$

$ab = 18$

$b^2 = 9$

$b = 3, b > 0$

$a = 6$

15. $\frac{4}{5} = .8 =$ **0.8**

$\cos B = \frac{8}{10} = \frac{4}{5} = 0.8 = .8$

16. **(C)**

$x(1 + \frac{150}{100}) = 2.50x$

17. **(C)**

$P(\text{American Staffordshire Terrier} \mid \text{Male dog}) = \frac{6}{10} = 0.60$

18. **(A)**

$\cos(180° - x°) = -\cos(x) = -\frac{\sqrt{39}}{5}$.

19. **(C)**

The location of median is $\frac{8}{2} = 4$

The 4th number of road bridges among 7 US states is 14,307.

20. **(D)**

The distance traveled is $\frac{(1+3) \times 40}{2} = 80$.

21. **(A)**

$\frac{1}{3} = \frac{1}{3} \neq 1$

The system has no solution.

22. **(A)**

By using two points on the line, such as $(40, 25)$ and $(80, 30)$, the slope is $\frac{5}{40} = \frac{1}{8}$.

Module 2-Advanced

1. **(B)**

$g(f(x)) = (2x+1)^2 + 2(2x+1) - 1 = 4x^2 + 4x + 1 + 4x + 2 - 1 = 4x^2 + 8x + 2$

2. **(B)**

$V = x \cdot 4x \cdot 5x = 20x^3$

3. **28**

$3x + 8y = 150 \; x + y = 40 \rightarrow 3x + 3y = 120$

$5y = 30 \; y = 6, x = 34$

$x - y = 28$

4. **(A)**

$y = \frac{3}{x}$

$\frac{1}{y} = \frac{x}{3}$

5. **6**

$\frac{b+30}{2} = 28$

$a + 8 + 18 + 26 + 30 + 32 + 32 + 40 = 24 \times 8$.

$a = 6$

6. **(B)**

$A = \pi(5)^2 \frac{1}{6} - \frac{\sqrt{3}}{4} \times 25 = \frac{25\pi}{6} - \frac{25\sqrt{3}}{4}$

7. **(B)**

$$t = -\frac{b}{2a} = \frac{-15}{2(-4.9)} = 1.53$$

8. **(D)**

$$600(1 + 0.015)^x > 1,000$$

When $x = 35, 600(1.015)^{35} = 1010.328791$.

9. **(A)**

$$1,525 + 0.1d > 520 + 0.25d$$

$$1,005 > 0.15d$$

$$d < 6,700$$

10. **60**

$$0.5 \times 20 \times 12 - 20 \times 3 = 120 - 60 = 60$$

11. **(A)**

$$y = -\frac{1}{10}\left(x - \frac{1}{2}\right) + 4 = -\frac{1}{10}x + \frac{81}{20}.$$

$$20y = -2x + 81$$

$$2x + 20y = 81$$

12. **(C)**

Area of square B is 144 square inches. The length of square B is 12 inches.

13. **(C)**

The population mean weight of bags of chips is

$$9.5 < \mu < 10.1$$

9.6 is in the interval of possible value of mean.

14. **(B)**

The shortest distance is $\sqrt{100^2 + 80^2} = 128.06$

15. **(A)**

$$M + J + C = 12$$

$$J = M + 5$$

$$C = J - 3 = M + 2$$

$$3M = 5, M = \frac{5}{3}$$

$$\frac{\frac{5}{3}}{12} = \frac{5}{36} = 0.1388$$

16. **(B)**

If the data set include extremely higher outliers, then the median value is much smaller than the mean.

The higher outlier makes the mean move more toward the right skew.

17. **(B)**

$h = -\frac{b}{2a} = \frac{-b}{2(-1)} = 3$

$b = 6$

$f(3) = -9 + 18 + c = -14$

$c = -23$

$b + c = -17$

18. **(C)**

By the special right triangle ratio, $BD = 20$ and $CD = 20\sqrt{3}$.

$AC = \sqrt{50^2 + (20\sqrt{3})^2} = 60.827$

19. **(D)**

$f(2) = 2 \times (2)^2 + 1 \ f(3) = 2 \times 3^2 + 1 \ f(4) = 2 \times 4^2 + 1 \ f(n) = 2 \times n^2 + 1$

20. **(C)**

$2^{x+1} + 2^{x-1} = 2 \cdot 2^x + \frac{1}{2} \cdot 2^x = \frac{5}{2} \cdot 2^x = 5 \cdot 2^{x-1}$

21. **(D)**

(D) This statement is false because Box and whiskers plot does not indicate the number of data.

22. **1, 4**

$x^2 - 2x + (x - 2)^2 - 4(x - 2) - 4 = 0$

$x^2 - 2x + x^2 - 4x + 4 - 4x + 8 - 4 = 0$

$2x^2 - 10x + 8 = 0$

$x^2 - 5x + 4 = 0$

$x = 1, 4$

PRACTICE TEST 3

Math

22 QUESTIONS

- The questions in this section address a number of important math skills.

- Use a calculator is permitted for all questions.

- Reference

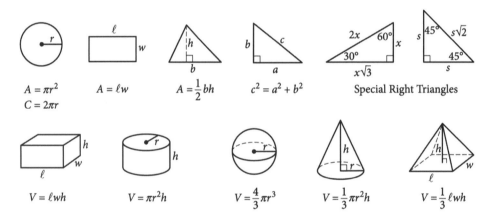

$A = \pi r^2$
$C = 2\pi r$

$A = \ell w$

$A = \frac{1}{2} bh$

$c^2 = a^2 + b^2$

Special Right Triangles

$V = \ell w h$

$V = \pi r^2 h$

$V = \frac{4}{3}\pi r^3$

$V = \frac{1}{3}\pi r^2 h$

$V = \frac{1}{3}\ell w h$

The number of degrees of arc in a circle is 360.

The number of radians of arc in a circle is 2π.

The sum of the measures in degrees of the angles of a triangle is 180.

For multiple-choice questions, solve each problem, choose the correct answer from the choices provided, and then circle your answer in this book. Circle only answer for each question. If you change your mind, completely erase the circle. You will not get credit for questions with more than one answer circled, or for questions with no answers circled.

For student-produced response questions, solve each problem and write your answer next to or under the question in the test book as described below.

- Once you've written your answer, circle it clearly. You will not receive credit for anything written outside the circle, or for any questions with more than one circled answer.

- **If you find more than one correct answer**, write and circle only one answer.

- Your answer can be up to 5 characters for a **positive** answer and up to 6 characters (Including the negative sign) for a **negative** answer, but no more.

- If your answer is a **fraction** that is too long (over 5 characters for positive, 6 characters for negative), write the decimal equivalent.

- If your answer is a **decimal** that is too long (over 5 characters for positive, 6 characters for negative), truncate it or round at the fourth digit.

- If your answer is a **mixed number** (such as $3\frac{1}{2}$), write it as an improper fraction (7/2) or its decimal equivalent (3.5).

- Don't include **symbols** such as a percent sign, comma, or dollar sign in your circled answer.

Answer	Acceptable ways to enter answer	Unacceptable: will NOT receive credit
3.5	3.5 3.50 7/2	31/2 31/2
$\frac{2}{3}$	2/3 .6666 .6667 0.666 0.667	0.66 .66 0.67 .67
$-\frac{1}{3}$	$-\frac{1}{3}$ $-.3333$ -0.333	$-.33$ -0.33

1. A company produces a snack package called Snacking. Each Snaking package contains exactly 2 container of yogurt, 1 cheese cube, and $\frac{1}{2}$ pounds of crackers. The company has in storage 600 containers of yogurt, 350 cheese cubes, and 100 pounds of crackers. Using only what the company has in storage, what is the maximum number of Snacking that can be produced?

   (A) 200

   (B) 300

   (C) 350

   (D) 400

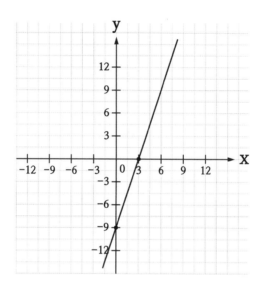

2. What is an equation of the line shown?

   (A) $3x - y = 9$

   (B) $3x + y = 9$

   (C) $3x + y = -9$

   (D) $x - y = 9$

$$P + 0.015P = 600$$

3. At the start of January, Jason deposited P dollars in a bank account. After the first deposit, Jason made no additional deposits or withdrawals. At the start of January of next year, Jason's deposit is increased to $600. Which of the following is the best interpretation of $0.015P$ in this context?

(A) The amount invested at the start of January

(B) The amount of yearly interest earned

(C) The total amount in the account at the start of January of next year

(D) The rate at which the investment earned interest during the year

$$\text{Circle A}: x^2 + y^2 = 25$$

$$\text{Circle B}: (x - 3)^2 + (y + 1)^2 = 25$$

4. In xy-plane, which translation of the graph of circle A would result in the graph of circle B?

(A) 3 units to the left and 1 unit up

(B) 3 units up and 1 unit to the left

(C) 3 units to the right and 1 unit down

(D) 3 units down and 1 unit to the left

5. The function f is defined by $f(x) = -100\,(2)^{2x} + 300$. What is the horizontal asymptote of $f(x)$?

	Volume (m^3)
Tetrahedron A	3.18
Tetrahedron B	25.44

6. Tetrahedron A is similar to tetrahedron B. The table gives the volumes, in cubic meters m^3, of the two tetrahedron. What is the ratio of side lengths, in meters, of tetrahedron A and B?

(A) $1 : 2$

(B) $1 : 4$

(C) $1 : 8$

(D) $8 : 1$

$$0.25x + 0.10y = 0.15\,(x + y)$$

7. The given equation represents a volume x, in liters, of a 25% saline solution that will be mixed with a volume y, in liters, of a 10% saline solution to produce a 15% saline solution. What volume in liters, of the 25% saline solution will be needed if 3 liters of the 10% saline solution is used?

8. Which of the following situation is best modeled by an exponential function?

(A) Each year, the number of people in a city decreases by 10% of the original number of the people in the city

(B) The number of people in a city decreases by 200 people per year.

(C) The number of people in a city is 400 fewer people at the end of every 2-year period.

(D) Each year, the number of people in a city decrease by 10% of the number of people in a city the preceding year

9. During the first x miles of 360 miles in North Dakota, she traveled an average of 60 miles per hour. For the rest of her trip in North Dakota, she traveled an average 20 miles per hour due to the rush hour. The total drive time in North Dakota was 8 hours. Which of the following equation represents this situation?

(A) $60 \left(360 - x\right) + 20x = \frac{1}{8}$

(B) $60x + 20 \left(360 - x\right) = \frac{1}{8}$

(C) $\frac{1}{60} \left(360 - x\right) + \frac{1}{20} \left(x\right) = 8$

(D) $\frac{1}{60}x + \frac{1}{20} \left(360 - x\right) = 8$

$$\frac{4 - x^2}{x + 2}$$

10. Which of the following is equivalent to the given expression?

(A) $x - 2$

(B) $2 - x$

(C) $\frac{x^2 - 4}{2 - x}$

(D) $-\frac{(x+4)(x-4)}{x+2}$

$$\sqrt{a} = \sqrt[3]{b}$$

$$a^{2x+1} = b$$

11. Two numbers, a and b, are each greater than zero. For what value of x does a and b satisfy the equations given above?

12. Class A and Class B each consist of 30 people. All the students in each class were asked to rate a quality of school cafeteria from 1 through 6. The results are summarized in the frequency table.

Rating	Frequency	
	Class A	Class B
1	1	2
2	5	3
3	6	10
4	10	5
5	4	6
6	4	4

Which statement correctly compares the Class A median rating to the Class B median rating?

(A) The Class A median rating is greater than the Class B median rating.

(B) The Class A median rating is smaller than the Class B median rating.

(C) The Class A median rating is equal to the Class B median rating.

(D) There is not enough information to compare the median ratings.

$$5x - 2y = 27$$

$$2x - 5y = -13$$

13. The solution to the given system of equations is (x, y). What is the value of $x - y$?

14. The population density of a city is 7690 people per square miles in the year 2020. The total area of the city is 80.8 square miles. Which of the following is closest to the population of the city in the year 2020 ?

(A) 95

(B) 616,000

(C) 621,000

(D) 6,210,000

15. Staff members of a high school newspaper want to estimate the average number of years teachers in the state have been teaching. At an annual conference for headteachers in the state, 50 headteachers were selected and asked how long they had been teaching. The survey showed that the average number of teaching years for teachers in the state is 15.5 years. Which of the following is true about the survey?

(A) It shows that all the teachers in the state have taught about 15.5 years.

(B) It shows that all the headteachers in the state have taught about 15.5 years.

(C) The survey sample should have included more conference attendees.

(D) The survey is biased because it is not representative of all teachers in the state.

16. The volume of a cone is 768 cubic inches, and the base of the cone has an area of 192 square inches. What is the height, in inches, of the cone?

(A) 4

(B) 12

(C) 14

(D) 16

$$20x^2 - 13x + 2 = 0$$

17. How many values of x satisfy the equation above?

(A) Zero

(B) One

(C) Two

(D) More than two

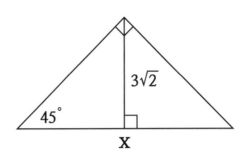

18. In the right triangle above, the opposite side of the right angle has length x. What is the value of x?

(A) 3

(B) $3\sqrt{2}$

(C) 6

(D) $6\sqrt{2}$

19.

$$qy - \frac{1}{2}x = \frac{2}{3} - y$$

$$-\frac{1}{2}x + 1 = qy - 4$$

In the given system of linear equations, q is the constant. If the system has no solution, what is the value of q?

20. What is the value of $\cos\left(\frac{3\pi}{4}\right)$?

(A) $-\frac{\sqrt{3}}{2}$

(B) $-\frac{\sqrt{2}}{2}$

(C) $\frac{\sqrt{3}}{2}$

(D) $\frac{\sqrt{2}}{2}$

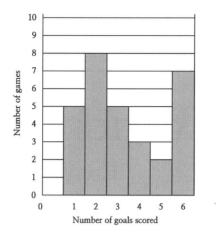

21. Based on the graph above, in how many of the games played did the soccer team score goals equal to the median number of goals for the 30 games?

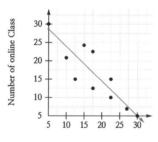

Number of on−campus Class

22. In a particular school district, 10 high schools were randomly selected for a study on the number of online courses and on-campus courses during December. The scatterplot above shows the number of courses surveyed by the administrators from 10 high schools. The line of best fit overestimates one high school's reported number of online courses by more than 5. For how many on-campus courses did the school report?

(A) 10

(B) 12.5

(C) 15

(D) 17.5

PRACTICE TEST 3
Easy

Math
22 QUESTIONS

- The questions in this section address a number of important math skills.

- Use a calculator is permitted for all questions.

- Reference

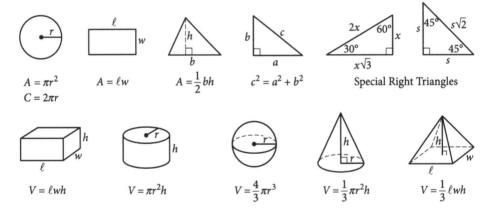

$A = \pi r^2$
$C = 2\pi r$

$A = \ell w$

$A = \frac{1}{2} bh$

$c^2 = a^2 + b^2$

Special Right Triangles

$V = \ell wh$

$V = \pi r^2 h$

$V = \frac{4}{3} \pi r^3$

$V = \frac{1}{3} \pi r^2 h$

$V = \frac{1}{3} \ell wh$

The number of degrees of arc in a circle is 360.

The number of radians of arc in a circle is 2π.

The sum of the measures in degrees of the angles of a triangle is 180.

For multiple-choice questions, solve each problem,choose the correct answer from the choices provided, and then circle your answer in this book. Circle only answer for each question. If you change your mind, completely erase the circle. You will not get credit for questions with more than one answer circled, or for questions with no answers circled.

For student-produced response questions, solve each problem and write your answer next to or under the question in the test book as described below.

- Once you've written your answer, circle it clearly. You will not receive credit for anything written outside the circle, or for any questions with more than one circled answer.

- **If you find more than one correct answer**, write and circle only one answer.

- Your answer can be up to 5 characters for a **positive** answer and up to 6 characters (Including the negative sign) for a **negative** answer, but no more.

- If your answer is a **fraction** that is too long (over 5 characters for positive, 6 characters for negative), write the decimal equivalent.

- If your answer is a **decimal** that is too long (over 5 characters for positive, 6 characters for negative), truncate it or round at the fourth digit.

- If your answer is a **mixed number** (such as $3\frac{1}{2}$), write it as an improper fraction (7/2) or its decimal equivalent (3.5).

- Don't include **symbols** such as a percent sign, comma, or dollar sign in your circled answer.

Answer	Acceptable ways to enter answer	Unacceptable: will NOT receive credit
3.5	3.5 3.50 7/2	31/2 31/2
$\frac{2}{3}$	2/3 .6666 .6667 0.666 0.667	0.66 .66 0.67 .67
$-\frac{1}{3}$	$-\frac{1}{3}$ $-.3333$ -0.333	$-.33$ -0.33

1. The function f is a linear function. The x-intercept of the graph of $y = f(x)$ in the xy-plane is $(4, 0)$. What is the x-intercept of the graph of $y = f(x + 2)$?

   (A) $(2, 0)$

   (B) $(6, 0)$

   (C) $(4, 2)$

   (D) $(4, -2)$

2. In the xy-plane, line ℓ has a slope of -3. Line k is perpendicular to the line ℓ and pass through the point $(-6, -5)$. Which of the following is an equation of line k?

   (A) $y = 3x + 13$

   (B) $y = 3x - 13$

   (C) $y = \frac{1}{3}x + 3$

   (D) $y = \frac{1}{3}x - 3$

3. In 2019, 15 Hydrangea were planted in a garden. The number of Hydrangea increased by 200% each year. Which of the following equations best models the estimated number of plants, P, in the garden t years after 2019?

   (A) $P(t) = 2(15)^t$

   (B) $P(t) = 3(15)^t$

   (C) $P(t) = 15(2)^t$

   (D) $P(t) = 15(3)^t$

4. In the xy-plane, the graph of $2x^2 - 8x + 2y^2 + 4y = 6$ is a circle. What is the radius of the circle?

   (A) $\sqrt{6}$

   (B) 3

   (C) $\sqrt{8}$

   (D) 4

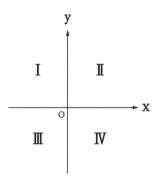

5. In the xy-plane shown, the quadrants are labeled I, II, III and IV. The graph of $y = -(x + h)^2 - k$, where $h > 0$ and $k > 0$, is a parabola. In which quadrant is the vertex of this parabola?

(A) Quadrant I

(B) Quadrant II

(C) Quadrant III

(D) Quadrant IV

6. Which of the following absolute value equations gives two solutions?

(A) $|x + 4| = 0$

(B) $2|x + 4| = -4$

(C) $|x + 4| - 4 = 3$

(D) $2|x + 4| + 2 = 1$

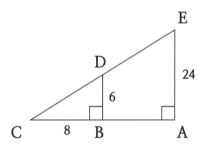

7. In the figure above $\overline{BD}$ is parallel to $\overline{AE}$. What is the perimeter of ACE?

8. Which of the following is equivalent to $\left(\frac{x}{2} + \frac{1}{x}\right)^2$

(A) $\frac{x^2}{2} + \frac{1}{x^2}$

(B) $\frac{x^2}{4} + \frac{1}{x^2}$

(C) $\frac{x^2}{4} + \frac{1}{x^2} + \frac{1}{2}$

(D) $\frac{x^2}{4} + \frac{1}{x^2} + 1$

9. $p\%$ of x is 5. Which expression represents x in terms of p?

(A) $\frac{5}{p}$

(B) $\frac{5p}{100}$

(C) $\frac{(100)(5)}{p}$

(D) $\frac{p}{(100)(5)}$

$$x + y = 12$$

$$x - y = 8$$

10. The solution to the given system of equations is (x, y). What is the value of $x^2 - y^2$?

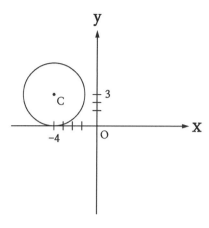

11. In the xy-plane above, the circle has center at $(-4, 3)$ and the circle is tangent to the x-axis. What is the equation of the circle $\odot C$?

(A) $(x - 4)^2 + (y - 3)^2 = 16$

(B) $(x + 4)^2 + (y - 3)^2 = 16$

(C) $(x - 4)^2 + (y - 3)^2 = 9$

(D) $(x + 4)^2 + (y - 3)^2 = 9$

12. What is the x-coordinate of the x-intercept of the line with equation $\frac{5}{4}x + \frac{3}{2}y = 2$ when it is graphed in the xy-plane?

$$y = -3$$

$$y = x^2 + kx + 5$$

13. In the system of equations shown, the system has exactly one solution. What is the possible value of k?

(A) $\sqrt{8}$

(B) $-\sqrt{8}, \ \sqrt{8}$

(C) $4\sqrt{2}$

(D) $-4\sqrt{2}, \ 4\sqrt{2}$

14. A National Air and Space museum built a scale model of the solar system throughout the building, where 1 mile in the model represents an actual distance of 300,000,000 miles. The model of the Sun is x miles away from the model of Earth. Which expression represents the actual distance, in thousands miles, between Earth and the Sun?

(A) $300,000,000x$

(B) $300,000x$

(C) $300x$

(D) $\frac{x}{300}$

15. The table summarizes the number of public schools in two California counties in 2018.

School	County	
	San Diego	Los Angeles
Elementary	498	1,395
Middle	165	422
High	191	570

A public high school will be selected at random from the two counties. What is the probability, to the nearest hundredth, of selecting a school in Los Angeles?

(A) 0.18

(B) 0.24

(C) 0.25

(D) 0.75

Time (hours)					
Abby	4.2	5	7	8	4.3
Brenda	3.7	3.8	4.6	4.4	x

16. Abby and Brenda each recorded the number of hours spent on a cellphone over 5 days. The number of hours spent on a cellphone is shown in the table. The average number of hours of phone usage for Abby was 0.7 hours greater than the average number of hours of phone usage for Brenda. What is the value of x?

17. Justin deposited x dollars in his investment account on January 1, 1990. The amount of money in the account triples each year until Justin had $1,215 in his investment account on January 1, 1994. What is the value of x?

(A) 15

(B) 45

(C) 135

(D) 405

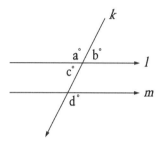

18. In the figure, lines ℓ and m each intersect line k. Which of the following is sufficient to prove that lines ℓ and m are parallel?

(A) $a° = b°$

(B) $a° = c°$

(C) $b° = c°$

(D) $a° = d°$

19. If $2(5x - 20) + 7(5x - 20) = 81$, what is the value of $5x$?

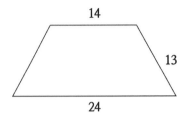

14

13

24

20. A painter charges \$2.5 per square foot to paint the top surface of a wooden roof. What would be the painter's charge for painting the top surface of the isosceles trapezoid wooden roof shown above?

(A) \$394.50

(B) \$570.00

(C) \$617.50

(D) \$1140.00

21. Point A and B lie on a circle with radius 2, and arc AB has length $\frac{2\pi}{3}$. What fraction of the circumference of the circle is the length of arc AB ?

22. What is the product of the complex numbers $2 + 3i$ and $2 - 3i$, where $i = \sqrt{-1}$?

(A) -5

(B) 13

(C) $4 - 9i$

(D) $4 + 9i$

PRACTICE TEST 3
Advanced

Math
22 QUESTIONS

- The questions in this section address a number of important math skills.

- Use a calculator is permitted for all questions.

- Reference

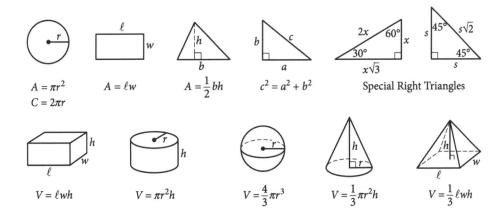

$A = \pi r^2$
$C = 2\pi r$

$A = \ell w$

$A = \frac{1}{2}bh$

$c^2 = a^2 + b^2$

Special Right Triangles

$V = \ell wh$

$V = \pi r^2 h$

$V = \frac{4}{3}\pi r^3$

$V = \frac{1}{3}\pi r^2 h$

$V = \frac{1}{3}\ell wh$

The number of degrees of arc in a circle is 360.

The number of radians of arc in a circle is 2π.

The sum of the measures in degrees of the angles of a triangle is 180.

For multiple-choice questions, solve each problem, choose the correct answer from the choices provided, and then circle your answer in this book. Circle only answer for each question. If you change your mind, completely erase the circle. You will not get credit for questions with more than one answer circled, or for questions with no answers circled.

For student-produced response questions, solve each problem and write your answer next to or under the question in the test book as described below.

- Once you've written your answer, circle it clearly. You will not receive credit for anything written outside the circle, or for any questions with more than one circled answer.

- **If you find more than one correct answer**, write and circle only one answer.

- Your answer can be up to 5 characters for a **positive** answer and up to 6 characters (Including the negative sign) for a **negative** answer, but no more.

- If your answer is a **fraction** that is too long (over 5 characters for positive, 6 characters for negative), write the decimal equivalent.

- If your answer is a **decimal** that is too long (over 5 characters for positive, 6 characters for negative), truncate it or round at the fourth digit.

- If your answer is a **mixed number** (such as $3\frac{1}{2}$), write it as an improper fraction (7/2) or its decimal equivalent (3.5).

- Don't include **symbols** such as a percent sign, comma, or dollar sign in your circled answer.

Answer	Acceptable ways to enter answer	Unacceptable: will NOT receive credit
3.5	3.5 3.50 7/2	31/2 31/2
$\frac{2}{3}$	2/3 .6666 .6667 0.666 0.667	0.66 .66 0.67 .67
$-\frac{1}{3}$	$-\frac{1}{3}$ $-.3333$ -0.333	$-.33$ -0.33

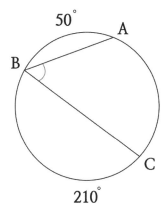

1. Points A, B, and C are on the circle shown above. The measure of arc BC is $210°$ and the measure of arc AB is $50°$. What is the measure of the inscribed angle, $\angle ABC$?

   (A) $40°$

   (B) $50°$

   (C) $100°$

   (D) $130°$

2. On a test, 5 multiple choices questions are asked. If each question has 4 choices and the person taking the test randomly guesses at the answers, what is the probability that this person will answer at least 1 question correctly?

   (A) $\frac{5}{1024}$

   (B) $\frac{243}{1024}$

   (C) $\frac{405}{1024}$

   (D) $\frac{781}{1024}$

3. If the graph of $f(x) = -x^3$ is translated 2 units left and 3 units up. The image of $f(x)$ represents $g(x)$. What is the value of $g(2.1)$?

   (A) -65.921

   (B) -13.81

   (C) 2.999

   (D) 2.99

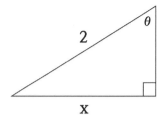

4. In the figure above, if $\sin\theta = \frac{x}{2}$,then what is the value of $\tan\theta$ in terms of x?

(A) $\frac{x}{\sqrt{2-x^2}}$

(B) $\frac{2}{\sqrt{4-x^2}}$

(C) $\frac{\sqrt{4-x^2}}{x}$

(D) $\frac{x}{\sqrt{4-x^2}}$

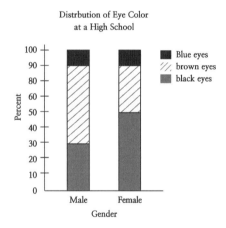

5. The segmented bar graph above shows the distribution of eye color of students at a high school by gender. There were 1000 male students and 300 more female students with black eyes than male students with black eyes. Which of the following best approximates the difference between the number of male students and female students?

(A) 200

(B) 300

(C) 400

(D) 500

6. A swimming pool is 3 feet deep in the shallow end. The bottom of the pool has a steady downward drop of 8° toward the deep end. If the pool is 50 feet long, how long is the deep end, to the nearest feet?

7. The fruit fly population in a certain laboratory doubles every week. Today there were 18 fruit flies. Which equation represents fruit fly population, $f(t)$, where t is the number of <u>months</u> from now?

(A) $f(t) = 18(2)^t$

(B) $f(t) = 18(2)^{4t}$

(C) $f(t) = 18(2)^{\frac{t}{4}}$

(D) $f(t) = 2(18)^{\frac{t}{4}}$

$$i + i^2 + i^3 + \ldots + i^{15}$$

8. Which of the following is equivalent to the expression given above when $i = \sqrt{-1}$?

(A) i

(B) $-i$

(C) 1

(D) -1

$$2x - y = 6$$

$$6x - 3y = 4k$$

9. For what value of k would the system of equations above have infinitely many solutions?

(A) 1.5

(B) 4.5

(C) 12

(D) 18

10. When a football team charges \$4 per ticket, average attendance is 400 people. For each \$1 increase in ticket price, average attendance decreases by 20 people. What is the ticket price to maximize revenue?

x	$f(x)$
0	1
2	$2a$
4	$4a^2$
6	$8a^3$

11. The table of function $f(x)$ is given. Which equation represents this relationship?

(A) $y = a(2)^x$

(B) $y = (2a)^x$

(C) $y = (2a)^{\frac{x}{2}}$

(D) $y = a(2)^{x-1}$

12. What are all values of x which $\sqrt{x^2 - 8}$ is a real number ?

(A) $|x| \geq \sqrt{8}$

(B) $|x| \leq \sqrt{8}$

(C) $|x| \geq 8$

(D) $|x| \leq 8$

13. A fire station is located 4 miles north and 6 miles east of Macy's plaza. There is a fire 20 miles south and 4 miles west of Macy's plaza. Which of the following is the distance, in miles, between the fire station and the fire?

(A) 20

(B) 22

(C) 24

(D) 26

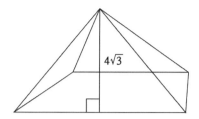

14. A pyramid with lateral faces of equilateral triangles is shown in the figure above. The slant height of the pyramid is $4\sqrt{3}$ inches. What is the volume, in cubic inches, of the pyramid?

    (A) 120.68

    (B) 147.80

    (C) 362.04

    (D) 443.41

$$27^{x^2} = 9^{5x-4}$$

15. What is the possible solution of the equation given above?

16. The average of 10 test scores is 80. When the highest score and lowest score are removed from the 10 scores, the average is 82. Which of the following is the average of the highest score and lowest score?

    (A) 10

    (B) 20

    (C) 72

    (D) 144

$$f(x) = \begin{cases} x^2 & for\ x > 2 \\ 3x - 1 & for\ x \leq 2 \end{cases}$$

17. The function $f(x)$ is given above. What is(are) the solution(s) of the piecewise function?

   (A) $(0,0)$

   (B) $\left(\frac{1}{3}, 0\right)$

   (C) No solution

   (D) Infinitely many solutions

18. The principal of Thomas Jefferson High School(TJHS) planned a science fair and conducted a study to estimate the percentage of the TJHS student population that received the Science and Engineering Awards. TJHS students in the science fair were asked about their awards, and their responses were recorded. Students who have not involved in the fair were NOT included in the study. Which of the following phrases best describes the principal's study?

   (A) Randomized experiment

   (B) Nonrandomized experiment

   (C) Randomized sample survey

   (D) Nonrandomized sample survey

19. For all x such that $0 \leq x \leq \frac{\pi}{2}$, which of the following expression is NOT equal to the $\cos x$?

   (A) $\cos(-x)$

   (B) $\sin\left(\frac{\pi}{2} - x\right)$

   (C) $\sin\left(x + \frac{\pi}{2}\right)$

   (D) $\cos(\pi - x)$

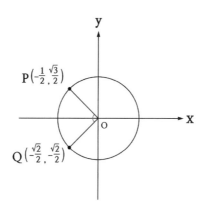

20. In the figure shown, points P and Q lie on a unit circle in the xy-plane, where O is the center of the circle. The measure, in radians, of angle POQ is $n\pi$. What is the value of n?

21. What is the remainder When the polynomial $f(x) = x^3 - 5x^2 - 7x + 12$ is divided by $x + 2$?

    (A) -14

    (B) -2

    (C) 0

    (D) 12

22. Linda ran a 6 mile cross-country course in 42 minutes while Monica ran the same course in 48 minutes. What is the difference between Linda's average speed and Monica's average speed, in <u>miles per hour</u>,to the nearest tenth?

Practice Test 3 Answers

\multicolumn{8}{c}{Practice Test 3-Module 1 ANSWERS}
Number	Answer	Number	Answer	Number	Answer	Number	Answer
1	A	7	1.5	13	2	19	$-\frac{1}{2} = -0.5 = -.5$
2	A	8	D	14	C	20	B
3	B	9	D	15	D	21	5
4	C	10	B	16	B	22	B
5	300	11	$\frac{1}{4} = .25 = 0.25$	17	C		
6	A	12	A	18	D		

\multicolumn{8}{c}{Practice Test 3-Module 2-Easy ANSWERS}
Number	Answer	Number	Answer	Number	Answer	Number	Answer
1	A	7	96	13	D	19	29
2	D	8	D	14	B	20	B
3	D	9	C	15	D	21	$\frac{1}{6} = 0.166 = 0.167 = .1666 = .166$
4	C	10	96	16	8.5	22	B
5	C	11	D	17	A		
6	C	12	$\frac{8}{5} = 1.6$	18	D		

\multicolumn{8}{c}{Practice Test 3-Module 2-Advanced ANSWERS}
Number	Answer	Number	Answer	Number	Answer	Number	Answer
1	B	7	B	13	D	19	D
2	D	8	D	14	A	20	$\frac{7}{12} = .5833 = 0.583$
3	A	9	B	15	$\frac{4}{3}, 2$	21	B
4	D	10	12	16	C	22	1.1
5	A	11	C	17	B		
6	10	12	A	18	D		

Explanation
Module 1

1. **(A)**

   $\frac{600}{2} = 300, 350, \frac{100}{\frac{1}{2}} = 200$

   Snaking package contains 2 yogurt, 1 cheese cube, and $\frac{1}{2}$ pounds of crackers.

   Thus, the maximum number of Snaking is 200.

2. **(A)**

The graph passes through $(0, -9)$ and $(3, 0)$.

(A) x-intercept is $(3, 0)$ and y-intercept is $(0, -9)$.

3. **(B)**

$0.015P$ represents the total amount of annual interest.

4. **(C)**

The center of circle A is $(0, 0)$, and the center of circle B is $(3, -1)$. The center is shifted 3 units to the right and 1 unit down from A to B.

5. **300**

The horizontal asymptote is the vertical shift, that is 300.

6. **(A)**

The ratio of volumes is $3.18 : 25.44 = 1 : 8$ The ratio of lengths is $1 : 2$

7. **$1.5 = \frac{3}{2}$**

$0.25x + 0.3 = 0.15(x + 3)$

$0.1x = 0.15$

$x = 1.5$

8. **(D)**

(D) The decay factor is 0.9

9. **(D)**

The time for x miles is $\frac{x}{60}$ hours. The time for $360 - x$ miles is $\frac{360-x}{20}$ hours.

$\frac{x}{60} + \frac{360-x}{20} = 8$

10. **(B)**

$\frac{-(x^2-4)}{x+2} = \frac{-(x-2)(x+2)}{x+2} = -x + 2$

11. **$\frac{1}{4} = 0.25 = .25$**

$(a^{\frac{1}{2}})^3 = (b^{\frac{1}{3}})^3$

$a^{\frac{3}{2}} = a^{2x+1}$

$2x + 1 = \frac{3}{2}$

$x = \frac{1}{4}$

12. **(A)**

The location of median is 15.5, that is, the median is the average of 15th and 16th values.

Median of class A is 4 and median of class B is 3.5

13. **2**

$$7x - 7y = 14$$

$$x - y = 2$$

14. **(C)**

Let N be the population of the city in 2020.

Density=$7690 = \frac{N}{80.8}$ $N = 621,352 \approx 621,000$

15. **(D)**

The population of the survey is teachers in the state, but selected sample contains only headteachers in the state.

The sample is not randomly selected.

16. **(B)**

$$768 = \frac{1}{3}Bh = \frac{1}{3}192h$$

$$h = 12$$

17. **(C)**

$$D = 13^2 - 4(20)(2) = 9 > 0$$

The quadratic equation has two zeros.

18. **(D)**

By the special triangle ratio, the opposite side of angle $45°$ is $3\sqrt{2}$. Then, $x = 2\times(3\sqrt{2}) = 6\sqrt{2}$.

19. $-\frac{1}{2} = -0.5 = -.5$

$$-\frac{1}{2}x + (1+q)y = \frac{2}{3}$$

$$-\frac{1}{2}x - qy = -5$$

$$\frac{-\frac{1}{2}}{-\frac{1}{2}} = \frac{1+q}{-q}$$

$$1 + q = -q$$

$$q = -\frac{1}{2}$$

20. **(B)**

$$\cos\left(\frac{3\pi}{4}\right) = -\frac{\sqrt{2}}{2}$$

21. **5**

The median number of goals scored is 3 for 30 games. The number of games played is 5.

22. **(B)**

The number of online class for the school that a line of best fit overestimate more than 5 is 15.

For the school, the number of on-campus class is 12.5

Module 2-Easy

1. **(A)**

   $y = f(x + 2)$ represents the horizontal translation 2 units to the left.

   $(4, 0) \rightarrow (2, 0)$

2. **(D)**

   $y = \frac{1}{3}(x + 6) - 5$

   $y = \frac{1}{3}x - 3$

3. **(D)**

   $P(t) = 15(1 + \frac{200}{100})^t = 15(3)^t$

4. **(C)**

   $x^2 - 4x + y^2 + 2y = 3$

   $(x^2 - 4x + 4) + (y^2 + 2y + 1) = 8$

   $r = \sqrt{8}$

5. **(C)**

   The vertex is $(-h, -k)$, where $-h < 0$ and $-k < 0$.

   Since the x-coordinate and y-coordinate of the vertex are negative, the vertex is in the third quadrant.

6. **(C)**

   (C) $|x + 4| = 7$

   Since 7 is positive, there are two solutions.

7. **96**

   The ratio of lengths of two triangles is $1 : 4$.

   The perimeter of triangle CBD is 24.

   $1 : 4 = 24 : P$

   $P = 96$

8. **(D)**

   $(\frac{x}{2} + \frac{1}{x})^2 = \frac{x^2}{4} + 1 + \frac{1}{x^2}$

9. **(C)**

   $\frac{p}{100} \times x = 5$

   $x = \frac{500}{p}$

10. **96**

    $x^2 - y^2 = (x - y)(x + y) = 12 \times 8 = 96$

11. **(D)**

    The center of the circle is $(-4, 3)$ and the radius is 3.

    $(x + 4)^2 + (y - 3)^2 = 9$

12. $\frac{8}{5} = \mathbf{1.6}$

    $\frac{5}{4}x = 2$ $x = \frac{8}{5}$

13. **(D)**

    $x^2 + kx + 5 = -3$

    $x^2 + kx + 8 = 0$

    $k^2 - 4(8) = 0$

    $k = \pm 4\sqrt{2}$

14. **(B)**

    Let y be the actual distance, in thousands miles, between Earth and the Sun.

    $1 : 300 \times 10^6 = x : y$

    $y = 300,000,000x = 300,000x$ thousands miles

15. **(D)**

    $P(\text{Los Angeles} \mid \text{High School}) = \frac{570}{190 + 570} = 0.749$

16. **8.5**

The total hours of phone usage for Abby is $4.2 + 5 + 7 + 8 + 4.3 = 28.5$

The total hours of phone usage for Brenda is $3.7 + 3.8 + 4.6 + 4.4 + x = 16.5 + x$

$16.5 + x + 0.7 \times 5 = 28.5$

$x = 8.5$

17. **(A)**

$x(3)^4 = 1,215 \ x = 15$

18. **(D)**

(D) If alternate exterior angles such as $a° = d°$ are congruent, two lines are parallel.

19. **29**

$9(5x - 20) = 81$

$5x - 20 = 9$

$5x = 29$

20. **(B)**

Area of trapezoid is $\frac{(14+24)12}{2} = 228$.

The painter's charge for painting the roof is $228 \times 2.5 = 570$.

21. $\frac{1}{6} = \mathbf{0.166 = 0.167 = .1666 = .1667}$

$\frac{\frac{2\pi}{3}}{4\pi} = \frac{1}{6}$

22. **(B)**

$(2 + 3i)(2 - 3i) = 2^2 + 3^2 = 4 + 9 = 13$

Module 2-Advanced

1. **(B)**

The measure of arc AC is $360 - 260 = 100°$.

The measure of the inscribed angle is $\frac{1}{2} \times 100 = 50°$

2. **(D)**

P(At least 1 correct question)=1-P(No correct question)=$1 - (\frac{3}{4})^5 = \frac{781}{1024}$

3. **(A)**

$g(x) = -(x+2)^3 + 3$

$g(2.1) = -(4.1)^3 + 3 = -65.921.$

4. **(D)**

The adjacent side of θ is $\sqrt{4 - x^2}$.

$\tan \theta = \frac{x}{\sqrt{4-x^2}}$

5. **(A)**

The number of male students with black eyes is $1000(0.3) = 300$

The number of female students with black eyes is 600.

$600 = 0.5 \times F$

$F = 1,200$

The difference between male students and female students is 200.

6. **10**

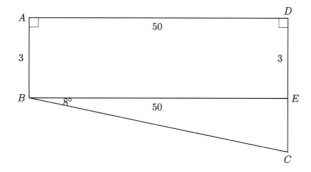

$EC = 50 \cdot \tan 8° = 7.027$

$CD = 3 + 7.027 \approx 10$

7. **(B)**

1 week $= \frac{1}{4}$ month.

$f(t) = 18 \times (2)^{4t}$

8. **(D)**

$i + i^2 + i^3 + i^4 = 0$

$i^{13} + i^{14} + i^{15} = i + i^2 + i^3 = -1$

9. **(B)**

$$\frac{2}{6} = \frac{1}{3} = \frac{6}{4k}$$

$$k = \frac{9}{2} = 4.5$$

10. **12**

$$R(x) = (4+x)(400-20x)$$

The maximum revenue occurs at $x = \frac{-4+20}{2} = 8$.

The ticket price when $x = 8$ is \$12.

11. **(C)**

The table represents the exponential relationship.

The initial value is $(0,1)$.

The growth factor is $2a$ every 2 increases in x.

$$y = (2a)^{\frac{x}{2}}$$

12. **(A)**

$$x^2 - 8 \geq 0$$

$$(x - \sqrt{8})(x + \sqrt{8}) \geq 0$$

$$|x| \geq \sqrt{8}$$

13. **(D)**

The distance between the fire station and the fire is $\sqrt{10^2 + 24^2} = \sqrt{676} = 26$

14. **(A)**

The length of base is 8. The area of base is 64.

The height of pyramid is $\sqrt{(4\sqrt{3})^2 - 4^2} = \sqrt{32}$.

$$V = \frac{1}{3} \times 64 \times \sqrt{32} = 120.679.$$

15. **$2, \frac{4}{3}$**

$$3^{3x^2} = 3^{2(5x-4)}$$

$$3x^2 = 10x - 8$$

$$x = 2, \frac{4}{3}$$

16. **(C)**

$\bar{x} = \frac{x_1+x_2+x_3+\cdots+x_{10}}{10} = 80$

$800 - (x_1 + x_{10}) = 82 \times 8$

$\frac{x_1+x_{10}}{2} = \frac{144}{2} = 72$

17. **(B)**

There is a solution at $(\frac{1}{3}, 0)$.

When $x \leq 2$, $3x - 1 = 0$

$x = \frac{1}{3}$

18. **(D)**

This study is a sample survey as the purpose is to collect information on the number of students who received the Science and Engineering Awards. However, the selected students are from the science fair and are not representative of all students in the school. Therefore, the study is a nonrandomized sample study.

19. **(D)**

(A) $\cos(-x) = \cos(x)$

(B) $\sin(\frac{\pi}{2} - x) = \cos(x)$

(C) $\sin(x + \frac{\pi}{2}) = \cos(-x) = \cos(x)$

(D) $\pi - x$ is in the second quadrant. $\cos(\pi - x) = -\cos(x)$

20. $\frac{7}{12} = .5833 = \mathbf{0.583}$

Let $\alpha + \beta$ be $\angle POQ$.

$\sin \alpha = \frac{\sqrt{3}}{2}$

$\sin \beta = \frac{\sqrt{2}}{2}$

$\alpha = \frac{\pi}{3}$

$\beta = \frac{\pi}{4}$

$\angle POQ = \frac{7}{12}\pi$

21. **(B)**

By the remainder theorem, $f(-2) = -8 - 5(-4) + 14 + 12 = -28 + 26 = -2$

22. **1.1**

Linda's average speed is $\frac{6}{\frac{42}{60}} = 8.5714$.

Monica's average speed is $\frac{6}{\frac{48}{60}} = 7.5$

The difference is $1.0714 \approx 1.1$

PRACTICE TEST 4

Math

22 QUESTIONS

- The questions in this section address a number of important math skills.

- Use a calculator is permitted for all questions.

- Reference

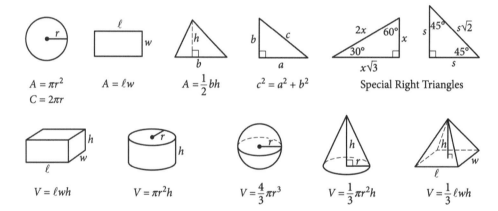

$A = \pi r^2$
$C = 2\pi r$

$A = \ell w$

$A = \frac{1}{2}bh$

$c^2 = a^2 + b^2$

Special Right Triangles

$V = \ell wh$

$V = \pi r^2 h$

$V = \frac{4}{3}\pi r^3$

$V = \frac{1}{3}\pi r^2 h$

$V = \frac{1}{3}\ell wh$

The number of degrees of arc in a circle is 360.

The number of radians of arc in a circle is 2π.

The sum of the measures in degrees of the angles of a triangle is 180.

For multiple-choice questions, solve each problem,choose the correct answer from the choices provided, and then circle your answer in this book. Circle only answer for each question. If you change your mind, completely erase the circle. You will not get credit for questions with more than one answer circled, or for questions with no answers circled.

For student-produced response questions, solve each problem and write your answer next to or under the question in the test book as described below.

- Once you've written your answer, circle it clearly. You will not receive credit for anything written outside the circle, or for any questions with more than one circled answer.

- **If you find more than one correct answer**, write and circle only one answer.

- Your answer can be up to 5 characters for a **positive** answer and up to 6 characters (Including the negative sign) for a **negative** answer, but no more.

- If your answer is a **fraction** that is too long (over 5 characters for positive, 6 characters for negative), write the decimal equivalent.

- If your answer is a **decimal** that is too long (over 5 characters for positive, 6 characters for negative), truncate it or round at the fourth digit.

- If your answer is a **mixed number** (such as $3\frac{1}{2}$), write it as an improper fraction (7/2) or its decimal equivalent (3.5).

- Don't include **symbols** such as a percent sign, comma, or dollar sign in your circled answer.

Answer	Acceptable ways to enter answer	Unacceptable: will NOT receive credit
3.5	3.5 3.50 7/2	31/2 31/2
$\frac{2}{3}$	2/3 .6666 .6667 0.666 0.667	0.66 .66 0.67 .67
$-\frac{1}{3}$	$-\frac{1}{3}$ $-.3333$ -0.333	$-.33$ -0.33

1.

$$4x^3 + bx^2 - 64x - 16b$$

In the polynomial above, b is constant. Which of the following is a factor of the polynomial?

(A) $x + 4$

(B) $x + b$

(C) $4x^3 + b$

(D) $x^2 + 16$

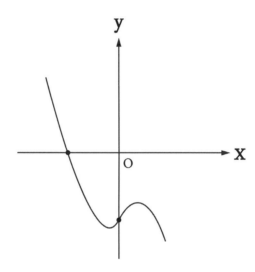

2. Which of the following could be equation of the graph in the xy-plane above?

(A) $y = -2\left(x^2 + 1\right)(x + 3)$

(B) $y = 2\left(x^2 + 1\right)(x + 3)$

(C) $y = -2\left(x^2 + 1\right)(x - 3)$

(D) $y = 2(x^2 + 1)(x - 3)$

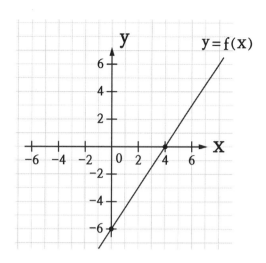

3. The graph of the linear function f is shown in the xy-plane above. What is the equation of a line that is perpendicular to the graph of f and passes through the point $(6, 0)$?

(A) $2x - 3y = -12$

(B) $2x - 3y = 12$

(C) $2x + 3y = 12$

(D) $2x + 3y = -12$

4. The gas tank in Sean's car has a fuel capacity of 18 gallons of gasoline, and the car averages 21 miles per gallon. If Sean has driven m miles since filling the gas tank, which of the following expressions represents the number of gallons of gasoline remaining in Sean's gas tank?

(A) $18 - 21m$

(B) $18 - \frac{m}{21}$

(C) $21m - 18$

(D) $\frac{m}{21} - 18$

5. A baker is making cakes. Each cake requires 3 eggs and a dozen of eggs costs $8. If the baker plans to make 100 cakes, what is the cost for eggs needed for 100 cakes?

$$P(b) = 3.50(b - 1,120)$$

6. Elsa sells b cups of coffee on average every month at her Café. The above equation gives Elsa's monthly profit $P(b)$, in dollars, for b cups of coffee sold. What is the meaning of the $1,120$ in the equation?

(A) Elsa will make $\$1,120$ profit if she sells 3.5 cups of coffee per day.

(B) Elsa must sell $1,120$ cups of coffee per month to break even with $\$0$ profit.

(C) Elsa must sell $1,120$ cups of coffee per month to maximize the profit.

(D) Elsa's monthly cost of running a Café is $\$1,120$.

7. On the first day of a new movie's release, a movie theater manager asks the people who saw the movie to rate it on a scale of 1 (worst) to 5 (best). The table below summarizes the responses of all 300 viewers of the movie.

	1	2	3	4	5	Total
Adults	15	40	72	35	38	200
Children	6	10	23	40	21	100
Total	21	50	95	75	59	300

What fraction of the adults survey gave a rating of 4 or 5 to the movie?

(A) $\frac{35}{200}$

(B) $\frac{73}{200}$

(C) $\frac{134}{300}$

(D) $\frac{73}{300}$

$$\sqrt{(x+2)^2} = 2$$

8. What is(are) the possible solution(s) x to the equation above?

    (A) $\{-2\}$

    (B) $\{0\}$

    (C) $\{-4, 2\}$

    (D) $\{0, -4\}$

$$h = -16t^2 + 80 + 96$$

9. A ball is thrown upward from the top of a $96 - $ ft tower with an initial velocity of 80 ft/sec. When does the ball reach its maximum height?

10. A piece of wire 48 inches is bent to form a rectangle. The length of the rectangle is 5 times the width. What is the length, in inches, of the rectangle?

    (A) 4

    (B) 8

    (C) 12

    (D) 20

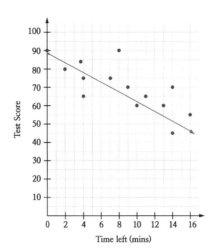

11. The scatterplot shows the relationship between the time, x, how many minutes left when a student turn in an exam and their exam scores, y, of 15 students. Which of the following equations models this relationship?

(A) $y = 85.7 + 2.14x$

(B) $y = 85.7 - 2.14x$

(C) $y = 85.7 + 0.214x$

(D) $y = 85.7 - 0.214x$

12. A circle in the xy-plane has the equation $(x-3)^2 + (y+5)^2 = 25$. Which of the following is the point that lies inside the circle?

(A) $(11, -5)$

(B) $(2, 2)$

(C) $(3, -3)$

(D) $(3, 0)$

13. A researcher studied the effects of taking an organic magnesium glycinate once a day on the bone health of older adults in Evergreen Senior Center in New Jersey. The sample is made up of 90 elderly adults in Evergreen Senior Center. The distribution of the ages of all the participants in the study is shown in the table below.

Age	Number of participants
$60 \leq X < 75$	8
$75 \leq X < 80$	12
$80 \leq X < 85$	29
$85 \leq X < 90$	33
$X \geq 90$	8
Total	90

What is the possible value of median participants of the study?

(A) 79

(B) 82

(C) 85

(D) 86

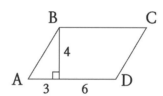

14. Parallelogram $ABCD$, with dimensions in inches, is shown in the diagram below. What is the perimeter of the parallelogram, in square inches?

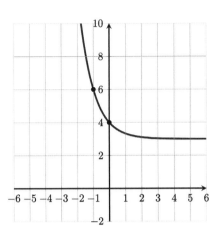

15. The graph of $y = f(x)$ is shown. What is the graph of $y = -f(x)$?

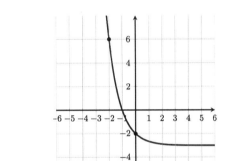

(A)

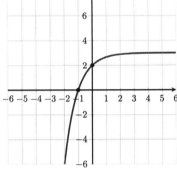

(B)

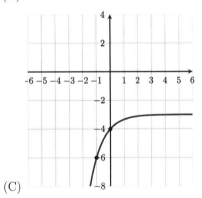

(C)

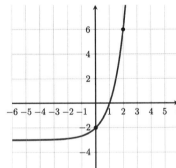

(D)

16. The number of rabbits in a forest is grown by about 3% each year for the next 15 years. If the population of rabbits this year is 100, which of the following of the equations models the population of rabbits, $R(t)$, in the forest t years from now?

(A) $R(t) = 100(0.03t)$

(B) $R(t) = 100(1.03t)$

(C) $R(t) = 100(1.03)^{\frac{t}{15}}$

(D) $R(t) = 100(1.03)^{t}$

17. A class A cargo ship can hold a maximum of 8430 standard containers. A class A cargo ship can hold 20% more containers than a class B cargo ship. How many more containers can a class of A cargo ship hold than a class B cargo ship?

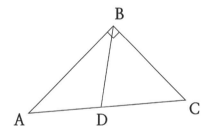

18. In triangle ABC above, point D lies on $\overline{AC}$. What is the value of $\sin(\angle ABD) - \cos(\angle CBD)$?

19. The senior class student council surveyed to determine whether the senior class would prefer a class trip to a museum. Among students surveyed, 55% preferred a museum and the survey's margin of error is 3%. Based on this information, which of the following is the most accurate conclusion?

(A) 55% of the senior students in the school prefer to visit a museum.

(B) Between 52% and 58% of all the students prefer to trip to a museum.

(C) Between 52% and 58% of senior students prefer to trip to a museum.

(D) There is insufficient information to determine whether more than half of senior students prefer to visit museums.

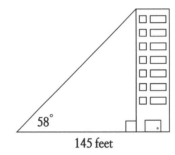

58°

145 feet

20. From a point on level ground 145 feet from the base of a tower, the angle of elevation to the top of the tower is 58°. How high is the tower, to the nearest feet ?

Employees in Washington (In thousands)

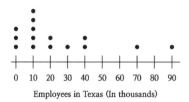

Employees in Texas (In thousands)

21. The dot plots show the number of employees in the industries, in thousands, of two states, Washington and Texas. Which of the following is(are) true?

I. The mean of the number of employees in Washington's industries is greater than the mean of the number of employees in Texas.

II. The standard deviation of the number of employees in Washington's industries is less than that of the number of employees in Texas.

(A) Neither I and II

(B) I only

(C) II only

(D) Both I and II

$$2x - y = 42$$

$$x + 2y = -4$$

22. If (x, y) is the solution to the system of equations shown above, what is the value of $x + y$?

(A) -10

(B) -6

(C) 6

(D) 16

PRACTICE TEST 4
Easy

Math
22 QUESTIONS

- The questions in this section address a number of important math skills.

- Use a calculator is permitted for all questions.

- Reference

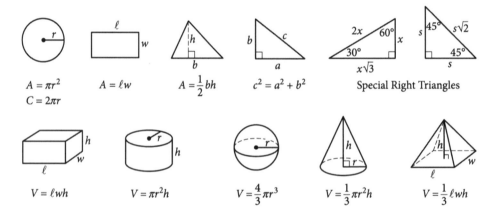

$A = \pi r^2$
$C = 2\pi r$

$A = \ell w$

$A = \frac{1}{2}bh$

$c^2 = a^2 + b^2$

Special Right Triangles

$V = \ell w h$

$V = \pi r^2 h$

$V = \frac{4}{3}\pi r^3$

$V = \frac{1}{3}\pi r^2 h$

$V = \frac{1}{3}\ell w h$

The number of degrees of arc in a circle is 360.

The number of radians of arc in a circle is 2π.

The sum of the measures in degrees of the angles of a triangle is 180.

For multiple-choice questions, solve each problem,choose the correct answer from the choices provided, and then circle your answer in this book. Circle only answer for each question. If you change your mind, completely erase the circle. You will not get credit for questions with more than one answer circled, or for questions with no answers circled.

For student-produced response questions, solve each problem and write your answer next to or under the question in the test book as described below.

- Once you've written your answer, circle it clearly. You will not receive credit for anything written outside the circle, or for any questions with more than one circled answer.

- **If you find more than one correct answer**, write and circle only one answer.

- Your answer can be up to 5 characters for a **positive** answer and up to 6 characters (Including the negative sign) for a **negative** answer, but no more.

- If your answer is a **fraction** that is too long (over 5 characters for positive, 6 characters for negative), write the decimal equivalent.

- If your answer is a **decimal** that is too long (over 5 characters for positive, 6 characters for negative), truncate it or round at the fourth digit.

- If your answer is a **mixed number** (such as $3\frac{1}{2}$), write it as an improper fraction (7/2) or its decimal equivalent (3.5).

- Don't include **symbols** such as a percent sign, comma, or dollar sign in your circled answer.

Answer	Acceptable ways to enter answer	Unacceptable: will **NOT** receive credit
3.5	3.5 3.50 7/2	31/2 31/2
$\frac{2}{3}$	2/3 .6666 .6667 0.666 0.667	0.66 .66 0.67 .67
$-\frac{1}{3}$	$-\frac{1}{3}$ $-.3333$ -0.333	$-.33$ -0.33

1. Which expression is equivalent to $a^{\frac{1}{8}} \left(a^{\frac{3}{4}} \right)^{\frac{3}{2}}$ where $a > 0$?

   (A) $\sqrt[4]{a^5}$

   (B) $\sqrt[3]{a^4}$

   (C) $\sqrt[8]{a^5}$

   (D) $\sqrt[8]{a^7}$

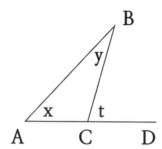

2. In the triangle ABC above, $\overline{CA}$ is extended to point D. Which of the following expresses t in terms of x and y?

   (A) $x + y$

   (B) $180 - x + y$

   (C) $-2x + y$

   (D) $180 - (x + y)$

$$3x - 5y = 20$$

$$ax + 15y = -60$$

3. In the system of equations above, a is a constant. If the system of equations has infinitely many solutions, what is the value of a ?

   (A) -9

   (B) -3

   (C) 3

   (D) 9

4. The graph of the linear function f in the xy-plane has a y-intercept of 12 and a slope of $-\frac{1}{5}$. What is the x-coordinate of x-intercept of f?

5. The speed of sound at sea level is approximately 761 miles per hour. What is the speed of sound at sea level measured in kilometers per second? (1 km = 0.621 miles)

(A) 0.340

(B) 20.424

(C) 471.96

(D) 1225.44

Population of Dubois, Wyoming	
Year	Population
2010	762
2020	746

6. The table above shows the population of Dubois, Wyoming for years 2010 and 2020. If the relationship between population and the year is linear, which of the following function P models the population of Dubois t years after 2010?

(A) $P(t) = 762 - 1.6t$

(B) $P(t) = 762 - 16t$

(C) $P(t) = 762 + 16(t - 2010)$

(D) $P(t) = 762 - 1.6(t - 2010)$

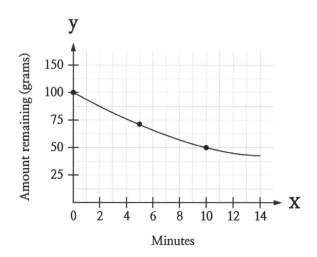

7. The graph of the function $f(x) = 100\left(\frac{1}{2}\right)^{\frac{x}{k}}$ shown in the xy-plane models the amount of a 100-gram sample of a radioisotope of Nitrogen-13 that would remain x minutes after the sample is obtained. In the function, k is a positive constant. What is the value of k?

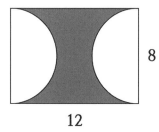

8. The figure shown represents a rectangle which two semicircles are cut out of. Which of the following gives the shaded area?

(A) 45.73

(B) 83.43

(C) 70.87

(D) 146.27

9. If the expression $\frac{x-15}{x^2+3x-28}$ is equivalent to $\frac{2}{x+7} + \frac{k}{x-4}$, what is the value of k?

(A) -2

(B) -1

(C) 1

(D) 2

10. A recipe for a cake suggests that a cook can substitute 1 tablespoon of margarine for $\frac{1}{4}$ cup of butter. The cheese cake calls for 2 cups of butter. How many cups of margarine could a cook substitute into the recipe? (16 tablespoon $=1$ cup)

(A) $\frac{1}{2}$

(B) 2

(C) 4

(D) 8

$$\frac{1}{y} = 6 + \frac{3xz}{75}$$

11. The equation above represents a reciprocal of y in terms of variables x and z. Which of the following equations correctly expresses y in terms of x and z ?

(A) $y = \frac{1}{6} + \frac{75}{3xz}$

(B) $y = \frac{150+xz}{25}$

(C) $y = \frac{25}{150+xz}$

(D) $y = \frac{1}{6+xz}$

12. The length of cube A is 6 times the length of cube B. The surface area of cube A is how many times the surface area of cube B?

$$121 - 100x^2 = 0$$

13. What are the solutions to the equation above?

(A) $x = \frac{100}{121}$

(B) $x = -\frac{100}{121}$ and $x = \frac{100}{121}$

(C) $x = \frac{11}{10}$

(D) $x = -\frac{11}{10}$ and $x = \frac{11}{10}$

Treatment	Number of plants		
	Withered	Survived	Total
Zero light	70	30	100
Low light	60	40	100
High light	15	85	100

14. The table above shows the results of an experiment involving the effect of amount of sunlight on plants. Based on the results, what fraction of the plants that survived received the zero light?

(A) 0.10

(B) 0.19

(C) 0.26

(D) 0.30

15. If $\left(x^4\right)^a \cdot \left(x^{\frac{1}{3}}\right)^2 = 1$, and $x > 1$, what is the value of a ?

(A) $-\frac{1}{6}$

(B) 0

(C) $\frac{1}{12}$

(D) $\frac{1}{6}$

16. Which of the following is an equation of the circle in the xy-plane that has endpoints of diameter at $(-4, 5)$ and $(8, -3)$?

(A) $(x - 4)^2 + (y - 2)^2 = \sqrt{52}$

(B) $(x - 4)^2 + (y - 2)^2 = 52$

(C) $(x - 2)^2 + (y - 1)^2 = \sqrt{52}$

(D) $(x - 2)^2 + (y - 1)^2 = 52$

17. A parabola in the xy-plane has equation $y = 2x^2 + 8x - 10$. Which equation shows the x-intercepts of the parabola as constants or coefficients?

(A) $y = 2\left(x^2 + 4x\right) - 10$

(B) $y = 2\left(x + 2\right)^2 - 18$

(C) $y = 2\left(x + 5\right)\left(x - 1\right)$

(D) $\frac{y + 18}{2} = (x + 2)^2$

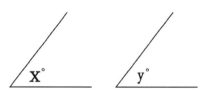

18. The angles shown above are acute and satisfy that $\sin(x^\circ) = \cos(y^\circ)$. If $x^\circ = 10t + 2$ and $y^\circ = 5t - 2$, what is the value of x° ?

Country	Median age of Population (years)
Russia	38.8
Germany	45.3
South Korea	32.5
Brazil	29.6
United States	37.1
Nigeria	17.9

19. The table given above represents the median ages of population of the countries in 1999. What is the range, in years, of the median ages of the populations for selected countries in the table?

20. Last year, Bethany's pear plants produced 48.5 kilograms of pears. This year, Bethany increased the number of pear plants in her garden by 32%. If her plants have pears this year at the same rate per plant, how many kilograms of pears can Bethany expect the plants to produce this year?

(A) 15.52

(B) 64.02

(C) 151.56

(D) 155.2

$$(2x + y)^2 - (2x - y)^2$$

21. The expression above is equivalent to $ax^2 + bxy + cy^2$, where a, b and c are constants. What is the value of b?

(A) 0

(B) 4

(C) 8

(D) 16

22. The party planner prepares a t cups of tea for p participants of the party, where $t = 3p + 10$ and $p > 1$. Based on the equation, how many cups of tea does the planner prepare for each additional participant in the party?

(A) 2

(B) 3

(C) 5

(D) 10

PRACTICE TEST 4
Advanced

Math
22 QUESTIONS

- The questions in this section address a number of important math skills.

- Use a calculator is permitted for all questions.

- Reference

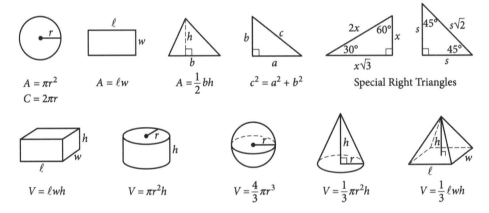

$A = \pi r^2$
$C = 2\pi r$

$A = \ell w$

$A = \frac{1}{2} bh$

$c^2 = a^2 + b^2$

Special Right Triangles

$V = \ell wh$

$V = \pi r^2 h$

$V = \frac{4}{3} \pi r^3$

$V = \frac{1}{3} \pi r^2 h$

$V = \frac{1}{3} \ell wh$

The number of degrees of arc in a circle is 360.

The number of radians of arc in a circle is 2π.

The sum of the measures in degrees of the angles of a triangle is 180.

For **multiple-choice questions**, solve each problem, choose the correct answer from the choices provided, and then circle your answer in this book. Circle only answer for each question. If you change your mind, completely erase the circle. You will not get credit for questions with more than one answer circled, or for questions with no answers circled.

For **student-produced response questions**, solve each problem and write your answer next to or under the question in the test book as described below.

- Once you've written your answer, circle it clearly. You will not receive credit for anything written outside the circle, or for any questions with more than one circled answer.

- **If you find more than one correct answer**, write and circle only one answer.

- Your answer can be up to 5 characters for a **positive** answer and up to 6 characters (Including the negative sign) for a **negative** answer, but no more.

- If your answer is a **fraction** that is too long (over 5 characters for positive, 6 characters for negative), write the decimal equivalent.

- If your answer is a **decimal** that is too long (over 5 characters for positive, 6 characters for negative), truncate it or round at the fourth digit.

- If your answer is a **mixed number** (such as $3\frac{1}{2}$), write it as an improper fraction (7/2) or its decimal equivalent (3.5).

- Don't include **symbols** such as a percent sign, comma, or dollar sign in your circled answer.

Answer	Acceptable ways to enter answer	Unacceptable: will NOT receive credit
3.5	3.5 3.50 7/2	31/2 31/2
$\frac{2}{3}$	2/3 .6666 .6667 0.666 0.667	0.66 .66 0.67 .67
$-\frac{1}{3}$	$-\frac{1}{3}$ $-.3333$ -0.333	$-.33$ -0.33

1. If the height of a right circular cylinder is increased by 10%, by what percent must the radius of the base be decreased so that the volume of the cylinder is decreased by 25% ?

    (A) 9%

    (B) 17%

    (C) 83%

    (D) 91%

2. Which of the following translations of the graph $y = x^2$ would transform to $y = x^2 - 4x + k$, where k is a constant greater than 5.

    (A) Left 2 units and up k units

    (B) Left 4 units and up $k - 4$ units

    (C) Right 2 units and up k units

    (D) Right 2 units and up $k - 4$ units

$$4x - 6 = ax - 6$$

3. In the given equation, a is a positive integer constant less that 5. The equation has exactly one solution. What is the greatest possible value of a ?

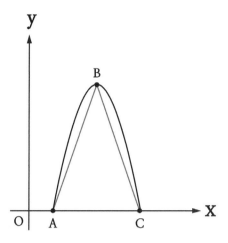

4. The quadratic function $f(x) = -2x^2 + 12x - 10$ and $\triangle ABC$ which is inscribed in $y = f(x)$ are graphed in the xy-plane above. Point B is at the maximum value of f. What is the area of $\triangle ABC$?

(A) 16

(B) 20

(C) 32

(D) 40

5. If a point (x, y) is chosen at random from the set of points where $-2 \le x \le 2$ and $-2 \le y \le 2$, what is the probability that the distance from the point (x, y) to the origin is less than or equal to 2?

(A) $\frac{1}{2}$

(B) $\frac{\pi}{4}$

(C) π

(D) 1

$$f(t) = 145 (1.015)^t$$

6. The function f, defined by the given equation, models the number of Fennec fox in Sahara Desert t years after 1900, where $0 \le t \le 40$. Which of the following equations best models the number of Fennec fox in Sahara Desert d decades after 1900? (1 decade=10 years)

(A) $f(d) = 145 \left(\frac{1.015}{10}\right)^d$

(B) $f(d) = 145 (1.015 \times 10)^d$

(C) $f(d) = 145 (1.015)^{\frac{d}{10}}$

(D) $f(d) = 145 (1.015)^{10d}$

7. Which of the following is the graph, in the xy-plane, of $y = \frac{2x^2 + 2x}{x}$?

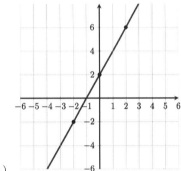

(A)

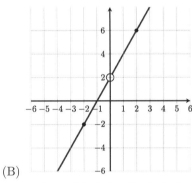

(B)

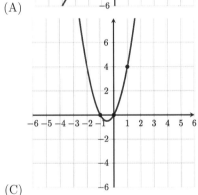

(C)

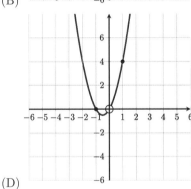

(D)

8. The New York Times has 1,620,000 online subscribers, of whom 580,000 are female, and 1,040,000 are male. Thirty percent of the females and fifty percent of the males renew the annual subscription contract. A random sample of 100 subscribers is selected. What is the expected number of subscribers who renew the contract?

(A) 30

(B) 43

(C) 57

(D) 80

$$2x^2 + 3x - 4 = 0$$

9. If one of the solutions of the given equation is $\frac{-3+\sqrt{t}}{4}$, where t is a constant, What is the value of t ?

10. For positive values of x and n, x is $n\%$ more than 20. Which expression represents x in terms of n?

(A) $20\left(\frac{n}{100}\right)$

(B) $20\left(\frac{100-n}{100}\right)$

(C) $20\left(\frac{100+n}{100}\right)$

(D) $20\left(\frac{n}{100}\right) + 1$

11. To check the effect of cold temperature on the friction of bike chains, 4 identical bike chains are lubricated with Dry lube, and then the bike chains are placed in a freezer for two hours. 4 identical bike chains are lubricated with Wet lube, and then the bike chains are kept at room temperature. The amount of friction each chain created is measured, and the mean for the cold chains is compared to the others. Is this a good experimental design?

(A) No, because more than 4 bike chains should be used.

(B) No, because the chains should be tested at more temperatures.

(C) No, because the temperature is confounded with types of chain lubricants

(D) Yes, because 2 treatments are tested in a well-designed setting.

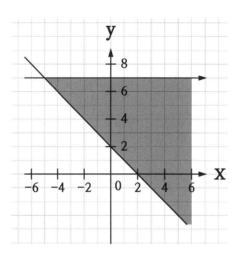

12. The solution to which system of inequalities is represented by the shaded region of the graph?

(A) $\begin{cases} y \le 7 \\ y \le -x + 2 \end{cases}$

(B) $\begin{cases} y \le 7 \\ y \ge -x + 2 \end{cases}$

(C) $\begin{cases} x \le 7 \\ y \le -x + 2 \end{cases}$

(D) $\begin{cases} x \le 7 \\ y \ge -x + 2 \end{cases}$

13. A function f has the property that if the point (a, b) is on the graph of the equation $y = f(x)$ in the xy-plane, then the point $(a + 1, \frac{1}{2}b)$ is also on the graph. Which of the following could define f?

(A) $f(x) = \frac{1}{2} \left(\frac{1}{12} \right)^x$

(B) $f(x) = 12 \left(\frac{1}{2} \right)^x$

(C) $f(x) = 12 (2)^x$

(D) $f(x) = \frac{1}{3} (12)^x$

14. A survey was conducted to determine what percentage of high school juniors would attend summer camps. In a random sample of 100 juniors, 34% indicated they would join the summer camp. The margin of error for the percentage of all juniors who would attend the summer camp is 7.8%. What is the possible percentage of all juniors who would attend the summer camp?

    (A) 34%

    (B) 26.2% to 41.8%

    (C) 30.1% to 37.9%

    (D) 30.6% to 37.4%

15. The Ferris wheel at a state fair has a radius of 50 feet, rotates at a constant speed, and completes 1 rotation in 5 minutes. How many degrees does the Ferris wheel rotate in 30 seconds?

    (A) 18°

    (B) 36°

    (C) 45°

    (D) 72°

16. Two numbers have a sum of 27 and a product of the two numbers is 81. What is sum of reciprocals of the two numbers?

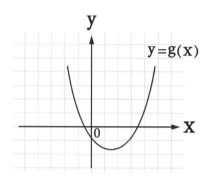

17. The graph of the function g is shown. Which of the following could define $y = -g(x)$?

    (A) $y = -x^2 - 2x + 1$

    (B) $y = -x^2 + 4x + 1$

    (C) $y = x^2 - 2x - 3$

    (D) $y = x^2 + 4x + 3$

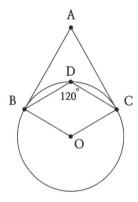

18. In the figure above, point O is the center of the circle, line segments AB and AC are tangent to the circle at points B and C, respectively, and the measure of $\angle BDC$ is 120°. What is the measure of $\angle BAC$?

19. Let m and s be the mean and standard deviation of the number of dogs and cats that several households raise in a particular city. If the number of pets each household raises would be half the current number of pets after 10 years, which of the following gives the mean and standard deviation of the number of dogs and cats of the same households in the city after 10 years?

(A) Mean=$\frac{m}{2}$ and Standard deviation=s

(B) Mean=m and Standard deviation=$\frac{s}{2}$

(C) Mean=$\frac{m}{2}$ and Standard deviation=$\frac{s}{2}$

(D) Mean=$\frac{m}{2}$ and Standard deviation=$\frac{s}{4}$

20. If the line $y = k$ is tangent to the circle $(x - 4)^2 + y^2 = 25$, what is one possible value of k?

21. A computer technician charges an initial fee of $50 and an additional $10 fee for every $\frac{1}{8}$ hour of work. The computer technician worked for h hours on a job, and the total cost for the job was $210. Which of the following equations models this situation?

(A) $50 + \left(10 \cdot \frac{h}{8}\right) = 210$

(B) $(50 + 10) \cdot (8h) = 210$

(C) $50 + (10 \cdot 8h) = 210$

(D) $(50 + 10)\left(\frac{1}{8}h\right) = 210$

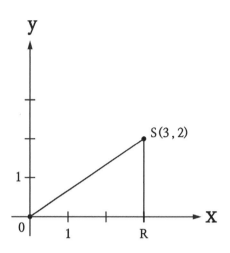

22. Triangle OSR in the figure above is revolved about the x-axis. What is the volume of the solid generated?

(A) 12.6

(B) 18.8

(C) 37.7

(D) 56.5

Practice Test 4 Answers

Practice Test 4-Module 1 ANSWERS							
Number	Answer	Number	Answer	Number	Answer	Number	Answer
1	A	7	B	13	B	19	C
2	A	8	D	14	28	20	232
3	C	9	$\frac{5}{2} = 2.5$	15	C	21	B
4	B	10	D	16	D	22	C
5	200	11	B	17	1405		
6	B	12	C	18	0		

Practice Test 4-Module 2-Easy ANSWERS							
Number	Answer	Number	Answer	Number	Answer	Number	Answer
1	A	7	10	13	D	19	27.4
2	A	8	A	14	B	20	B
3	A	9	B	15	A	21	C
4	60	10	A	16	D	22	B
5	A	11	C	17	C		
6	A	12	36	18	62		

Practice Test 4-Module 2-Advanced ANSWERS							
Number	Answer	Number	Answer	Number	Answer	Number	Answer
1	B	7	B	13	B	19	C
2	D	8	B	14	B	20	$-5, 5$
3	3	9	41	15	B	21	C
4	A	10	C	16	$\frac{1}{3} = 0.333 = .3333$	22	A
5	B	11	C	17	B		
6	D	12	B	18	60		

Explanation
Module 1

1. **(A)**

   $$4x^3 + bx^2 - 64x - 16b = x^2(4x + b) - 16(4x + b) = (x^2 - 16)(4x + b) = (x - 4)(x + 4)(4x + b)$$

2. **(A)**

   The graph has one negative zero, the highest degree must be an odd power, and its leading coefficient is negative.

   (A) The highest degree is 3, and it has zero at $x = -3$.

3. **(C)**

The slope of the given graph is $m = \frac{3}{2}$.

The slope of the perpendicular line is $\frac{-2}{3}$.

$y = -\frac{2}{3}(x - 6) = -\frac{2}{3}x + 4$

$2x + 3y = 12$

4. **(B)**

The amount of gallons of gasoline used to travel m miles is $\frac{m}{21}$. The amount of remaining gallons is $18 - \frac{m}{21}$.

5. **200**

$3 \times \frac{8}{12} \times 100 = 200$

6. **(B)**

The b-intercept is $(1,120,0)$.

When Elsa sells 1,120 cups of coffee per month, she can make break-even with a $0 profit.

7. **(B)**

$P(\text{adults} \mid 4 \text{ or } 5) = \frac{35+38}{200} = \frac{73}{200}$.

8. **(D)**

$\sqrt{(x+2)^2} = |x+2| = 2$

$x + 2 = \pm 2 \quad x = -4, 0$

9. $\frac{5}{2} = \mathbf{2.5}$

It hits the maximum height at the vertex.

$h = \frac{-b}{2a} = \frac{-80}{2(-16)=2.5} = \frac{5}{2}$

10. **(D)**

$l = 5w$

$p = 2(l + w) = 12w = 48 \quad w = 4, l = 20$

11. **(B)**

The slope of the line of best fit is negative.

(B) When $x = 10$, $y = 85.7 - 2.14(10) = 64.3$

(D) When $x = 10$, $y = 85.7 - 0.214(10) = 83.56$

12. **(C)**

The center of the circle is $(3, -5)$, and the radius is 5.

(C) $(3-3)^2 + (-3+5)^2 = 4 < 25$

The point $(3, -3)$ lies inside the circle.

13. **(B)**

The location of median is $\frac{90+1}{2} = 45.5$

The median lies in the age group, $80 \leq x < 85$.

14. **28**

$AB = \sqrt{3^2 + 4^2} = 5$

The perimeter of parallelogram is $5 + 5 + 9 + 9 = 28$.

15. **(C)**

For the graph of $y = -f(x)$, the horizontal asymptote is $y = -3$, y-intercept is $(0, -4)$, and $(-1, -6)$.

16. **(D)**

It is an exponential function as the growth factor is $1 + \frac{3}{100} = 1.03$.

$R(t) = 100(1.03)^t$

17. **1405**

$A = 8,430$

$8,430 = 1.2B \rightarrow B = 7,025$

$A - B = 1,405$

18. **0**

$\angle ABD + \angle CBD = 90°$

$\sin \angle ABD = \cos \angle CBD$

$\sin \angle ABD - \cos \angle CBD = 0$

19. **(C)**

The plausible value of percentage of people in the population who preferred a museum is $55\% \pm 3\%$.

(C) Between 52% and 58% of senior students prefer to trip to a museum.

20. **232**

$$\tan 58° = \frac{h}{145}$$

$$h = 232$$

21. **(B)**

I. This statement is true.

The mean number of employees in Washington is $\frac{540}{17} = 31.7647$.

The mean number of employees in Texas is $\frac{360}{15} = 24$.

II. This statement is false.

The dot plot of employees in Washington has two more outliers than the dot plot of employees in Texas, so the data of employees in Washington has a greater standard deviation.

22. **(C)**

$$4x - 2y = 84 \quad x + 2y = -4$$

$$5x = 80$$

$$x = 16, y = -10$$

$$x + y = 6$$

Module 2-Easy

1. **(A)**

$$a^{\frac{1}{8}}(a^{\frac{3}{4}})^{\frac{3}{2}} = a^{\frac{1}{8}} \cdot a^{\frac{9}{8}} = a^{\frac{10}{8}} = a^{\frac{5}{4}}$$

2. **(A)**

The exterior angle theorem is that the sum of remote interior angles is equal to the exterior angle.

$$t = x + y$$

3. **(A)**

$$\frac{3}{a} = \frac{-5}{15} = -\frac{1}{3}$$

$$a = -9$$

4. **60**

$$y = -\frac{1}{5}x + 12$$

The x-intercept is $(60, 0)$.

5. **(A)**

$$\frac{761}{0.621} \times \frac{1}{3,600} = 0.340$$

6. **(A)**

The slope is $m = \frac{746-762}{10} = -1.6$.

It passes through $(0, 762)$.

$$P(t) = 762 - 1.6t$$

7. **10**

The decay factor is $\frac{1}{2}$, and the decay period is 10.

$$k = 10$$

8. **(A)**

The shaded area $= 12 \times 8 - 16\pi = 45.73$

9. **(B)**

$$\frac{2}{x+7} + \frac{k}{x-4} = \frac{2(x-4)+k(x+7)}{(x-4)(x+7)} = \frac{x-15}{x^2+3x-28}$$

$$2 + k = 1$$

$$k = -1$$

10. **(A)**

The ratio of cups of margarine and butter is $\frac{1}{16} : \frac{1}{4} = x : 2$

$$x = \frac{1}{2}$$

11. **(C)**

$$\frac{1}{y} = \frac{450+3xz}{75}$$

$$y = \frac{75}{450+3xz} = \frac{25}{150+xz}$$

12. **36**

The ratio of lengths of A to B is $6 : 1$.

The ratio of surface areas of A to B is $36 : 1$.

13. **(D)**

$$x^2 = \frac{121}{100} \quad x = \pm\frac{11}{10}$$

14. **(B)**

$$P(\text{plants survived} \mid \text{zero light}) = \frac{30}{155} = 0.1935$$

15. **(A)**

$$x^{4a} \cdot x^{\frac{2}{3}} = x^{4a+\frac{2}{3}} = x^0$$

$$4a + \frac{2}{3} = 0$$

$$a = -\frac{1}{6}$$

16. **(D)**

The center is $\left(\frac{-4+8}{2}, \frac{5-3}{2}\right) = (2,1)$

$$(-4-2)^2 + (5-1)^2 = 36 + 16 = 52$$

$$(x-2)^2 + (y-1)^2 = 52$$

17. **(C)**

$$y = 2(x^2 + 4x - 5) = 2(x+5)(x-1)$$

18. **62**

$$x + y = 90$$

$$x + y = 10t + 2 + 5t - 2 = 15t = 90$$

$$t = 6$$

$$x = 62$$

19. **27.4**

The range is $45.3 - 17.9 = 27.4$

20. **(B)**

$$48.5(1.32) = 64.02$$

21. **(C)**

$$(2x+y)^2 - (2x-y)^2 = (2x+y+2x-y)(2x+y-2x+y) = (4x)(2y) = 8xy$$

22. **(B)**

$$m = \frac{\Delta t}{\Delta p} = 3$$

The increases in cups of tea is 3 for each additional participant in the party.

Module 2-Advanced

1. **(B)**

$$\left(1 - \frac{p}{100}\right)^2 \cdot (1.1) = 0.75$$

$$\left(1 - \frac{p}{100}\right) = 0.825722$$

$1 - 0.825722 = 0.174278$

$p\% = 17.4\%$

2. **(D)**

   $y = (x^2 - 4x + 4) + k - 4 = (x - 2)^2 + k - 4$

   The vertex of image is $(2, k - 4)$.

   This indicates that horizontal shift of 2 units to the right and vertical shift of $k - 4$ units up.

3. **3**

   $a \neq 4$

   $1 \leq a \leq 3$

   $a = 3$

4. **(A)**

   $f(x) = -2(x - 1)(x - 5) = -2(x - 3)^2 + 8$

   The x-intercept is $(1, 0)$ and $(5, 0)$.

   The vertex is $(3, 8)$.

   The area of triangle is $\frac{1}{2}(5 - 1)(8) = 16$

5. **(B)**

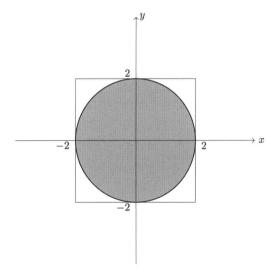

The points that are less than or equal to 2 from the origin are inside the circle of radius 2.

Probability $= \frac{\text{area of circle}}{\text{area of square}} = \frac{4\pi}{16} = \frac{\pi}{4}$.

6. **(D)**

1 year $= \frac{1}{10}$ decade.

The growth period is $\frac{1}{10}$.

$f(d) = 145(1.015)^{10d}$

7. **(B)**

The domain of the function is all real numbers except $x = 0$.

$y = \frac{x(2x+2)}{x} = 2x + 2$

The graph of the function is a line with a hole at $(0, 2)$.

8. **(B)**

$\frac{0.3 \times 580,000 + 0.5 \times 1,040,000}{1,620,000} \times 100 = 42.839 \approx 43$

9. **41**

$x = \frac{-3 \pm \sqrt{9 - 4(2)(-4)}}{4} = \frac{-2 \pm \sqrt{41}}{4}$

$t = 41$

10. **(C)**

$x = (1 + \frac{n}{100})20 = (\frac{100+n}{100})20$

11. **(C)**

It is not a well-designed experiment because we cannot compare three variables simultaneously. To compare the effect of temperature and the friction of bike chains, we must control the types of bike lubricants.

12. **(B)**

The equations of the graph are $y = 7$ and $y = -x + 2$.

The shaded region is above $y = -x + 2$, but below $y = 7$.

13. **(B)**

Since the decay factor is $\frac{1}{2}$, the relationship is an exponential function.

14. **(B)**

The possible percentage of all juniors who would attend the summer camp is $34\% \pm 7.8\% = 26.2\%$ to 41.8%.

15. **(B)**

$$\frac{360°}{5 \times 60} \times 30 = 36°$$

16. $\frac{1}{3} = \mathbf{0.333} = \mathbf{.3333}$

$x + y = 27$

$xy = 81$

$\frac{1}{x} + \frac{1}{y} = \frac{x+y}{xy} = \frac{27}{81} = \frac{1}{3}$

17. **(B)**

$y = -g(x)$ has a vertex at (h, k), where $h > 0, k > 0$, and it opens downward.

(B) $y = -x^2 + 4x + 1 = -(x-2)^2 + 5$

The vertex is $(2, 5)$

18. **60**

The measure of major arc BC is $240°$.

The measure of arc BDC is $120°$. Since $\triangle ABO$ and $\triangle ACO$ are right triangles and congruent, $m\angle BAO = m\angle CAO = 30°$

$m\angle BAC = 60°$.

19. **(C)**

Multiplying the random variables by $\frac{1}{2}$ changes the center and spread of the number of dogs and cats in the city 10 years later.

Thus, the new mean is $\frac{m}{2}$, and the new standard deviation is $\frac{s}{2}$.

20. $\mathbf{-5, 5}$

$y = k$ is a horizontal line that intersect with the circle exactly at one point.

Since the farthest y values from the center is $y = -5$ or $y = 5$, $k = \pm 5$.

21. **(C)**

The hourly payment is \$80.

$50 + 80x = 210$

22. **(A)**

$V = \frac{1}{3} \times \pi(2)^2 \times 3 = 4\pi = 12.566$

MEMO

MEMO